Quantum Gravity in a Nutshell1 Second Edition

Beyond Einstein, Volume 9

Balungi Francis

Published by Bill Stone Services, 2019.

Also by Balungi Francis

Beyond Einstein
Quantum Gravity in a Nutshell1
Solutions to the Unsolved Physics Problems
Mathematical Foundation of the Quantum Theory of Gravity
A New Approach to Quantum Gravity
Balungi's Approach to Quantum Gravity
QG: The strange theory of Space,Time and Matter
The Holy Grail of Modern Physics
Fifty Formulas that Changed the World
Quantum Gravity in a Nutshell1 Second Edition
What is Real?:Space Time Singularities or Quantum Black Holes?Dark
Matter or Planck Mass Particles? General Relativity or Quantum Gravity?
Volume or Area Entropy Law?
The Holy Grail of Modern Physics
Brief Solutions to the Big Problems in Physics, Astrophysics and Cosmology

Brief Solutions to the Big Problems
Brief Solutions to the Big Problems

Pursuing a Nobel Prize
Serious Scientific Answers to Millennium Physics Questions

Using Geographical Information Systems to Create an Agroclimatic Zone map for Soroti District

Think Physics
Proof of the Proton Radius
Emergence of Gravity
On the Deflection of Light in the Sun's Gravitational Field
Reinventing Gravity

Standalone
Using Gis to Create an Agroclimatic Zone map for Soroti Distric
Expecting
Quantum Gravity in a Nutshell 2
Balungi's Guide to a Healthy Pregnancy
Prove Physics
The Origin of Gravity and the Laws of Physics
Derivation of Newton's Law of Gravitation
When Gravity Breaks Down

Table of Contents

QUANTUM GRAVITY IN A NUTSHELL1

SECOND EDITION

BALUNGI FRANCIS

DEDICATION

To my wife W. Ritah for her constant feedback throughout and many long hours of editing,

To my sons Odhran and Leander,

To Carlo Rovelli, Lee Smolin, Neil deGrasse Tyson and Sabine Hossenfielder, I say thank you for your astonishing suggestions.

PREFACE

Today we are blessed with two extraordinarily successful theories of physics. The first is the General theory of relativity, which describe the large scale behavior of matter in a curved space time. This theory is the basis for the standard model of big bang cosmology. The discovery of gravitational waves at LIGO observatory in the US (and then Virgo, in Italy) is only the most recent of this theory's many triumphs.

The second is quantum mechanics. This theory describes the properties and behavior of matter and radiation at their smallest scales. It is the basis for the standard model of particle physics, which builds up all the visible constituents of the universe out of collections of quarks, electrons and force-carrying particles such as photons. The discovery of the Higgs boson at CERN in Geneva is only the most recent of this theory's many truimphs.

But, while they are both highly successful, those two structures leave a lot of important questions unanswered. They are also based on two different interpretations of space and time, and are therefore fundamentally incompatible. We have two descriptions but, as far as we know, we've only ever had one universe. What we need is a quantum theory of gravity.

There is a need for a book on a Quantum Theory of Gravity that is not directed at specialists but, rather, sets out the concepts underlying this subject for a broader scientific audience and conveys joy in their beauty. Balungi has written with this goal in mind, and has succeeded admirably. This wonderful and exciting book is optimal for physics graduate students and researchers. The physical explanations are exceedingly well written and integrated with formulas. Quantum Gravity is the next big thing and this book will help the reader understand and use the theory.

AUTHOR'S NOTE

Our search for ultimate understanding—the Quantum Theory of Gravity—has long been the quest of such great scientists as Aristotle, Newton, Einstein, Hawking and many others, and is expected to transform science, providing clarity and understanding that is unknown today, ideally via one single overlooked principle in nature. So far, this quest has produced theories such as Special Relativity, General Relativity and Quantum Mechanics, and such recent proposals as "Dark Matter" and "Dark Energy" in cosmology. Yet these all suffer serious internal problems and compatibility issues with each other, introducing even more questions, mysteries and paradoxes—and often even violations of our laws of physics upon closer examination. As a result, the Quantum Theory of Gravity continues to elude us, leaving a fractured and divided scientific community with no clear direction forward. This has also resulted in the mathematisation of physics which has resulted in the reduction of the cosmos to a mathematical entity, which has not only confused physicists but accounts for their worst and most distracting assertions. This book makes a first case for the latter, with clear discussions exposing the flaws in the above concepts and more, while stepping back to take a good look at the scientific legacy we have inherited.

We are probably asking the wrong questions at the moment, nevertheless it is impossible to resist the temptation to try. After all, the other fundamental forces – except gravity – fit very neatly with quantum mechanics.

Physics is an entity and therefore requires only one subject to describe it fully and this subject is quantum gravity.

Balungi Francis

PARTI FOUNDATIONS

1 GENESIS

The development of a quantum theory of gravity began in 1899 with Max Planck's formulation of "Planck scales" of mass, time, and length. During this period, the theories of quantum mechanics, quantum field theory and general relativity had not yet been developed. This means that Planck himself had no idea about what he had just developed-behind the Black board. Planck was not aware of quantum gravity and what it would mean for physicists. But he had just coined in formula one of the starting point for the holy grail of physics.

After P.Bridgman's disapproval of Planck's units in 1922, Albert Einstein having published the General Relativity theory, a few months after its publication he noted that "to the intra-atomic movement of electrons, atoms would have to radiate not only electromagnetic but also gravitational energy if only in tiny amounts, as this is hardly true in nature, it appears that quantum theory would have to modify not only Maxwellian electrodynamics, but also the new theory of gravitation". This showed Einstein's interest in the unification of Planck's quantum theory with his newly developed theory of Gravitation.

Then in 1933 came Bronstein's cGh-plan as we know it today. In his plan he argued a need for Quantum Gravity. In his own words he stated: "After the relativistic quantum theory is created, the task will be to develop the next part of our scheme that is, to unify quantum theory (h), special relativity (c) and the theory of gravitation (G) into a single theory". Thus the theory of quantum gravity is expected to be able to provide a satisfactory description of the microstructure of space time at the so called Planck scales, at which all fundamental constants of the ingredient theories, c (speed of light), h (Planck constant) and G (Newton's constant), come together to form units of mass, length and time.

The need for the theory of quantum gravity is crucial in understanding nature, from the smallest to the biggest particle ever known in the universe. For example, "we can describe the behavior of flowing water with the long- known classical theory of hydrodynamics, but if we advance to smaller and smaller scales and eventually come across individual atoms, it no longer applies. Then we need quantum physics just as a liquid consists of atoms". Daniel Oriti in this case imagines space to be made up of tiny cells or atoms of space and a new theory of quantum gravity is required to describe them fully.

The demand for consistency between a quantum description of matter and a geometric description of spacetime, as well as the appearance of singularities and the black hole information paradox indicate the need for a full theory of quantum gravity. For example; for a full description of the interior of black holes, and of the very early universe, a theory is required in which gravity and the associated geometry of space-time are described in the language of quantum physics. Despite major efforts, no complete and consistent theory of quantum gravity is currently known, even though a number of promising candidates exist.

For us to solve the problem of quantum gravity (QG) we need to address and understand in detail the situations where the general theory of relativity (GR) fails. That is; General relativity fails to account for dark matter, GR fails to explain details near or beyond space-time singularities. That is, for high or infinite densities where matter is enclosed in a very small volume of space. Abhay Ashtekar says that; when you reach the singularity in general relativity, physics just stops, the equations break down.

2GRAVITY

Gravity is the branch of physics relating to the very big. The histroy of gravity began with the following people; Archimedes and Vitruvius the architect both of Greco-Roman world, Aryabhata of ancient India and Galileo Galilei. Gravity is most accurately described by the general theory of relativity which describes gravity not as a force, but as a consequence of the curvature of spacetime caused by the uneven distribution of mass. The most extreme example of this curvature of spacetime is a black hole, from which nothing not even light can escape once past the black hole's event horizon.

However, for most applications, gravity is well approximated by Newton's law of universal gravitation, which describes gravity as a force which causes any two bodies to be attracted to each other, with the force proportional to the product of their masses and inversely proportional to the square of the distance between them.

Gravity is the weakest of the four fundamental forces of physics, as a consequence, it has no significant influence at the level of subatomic particles. In contrast, it is the dominant force at the macroscopic scale, and is the cause of the formation, shape and trajectory of astronomical bodies. For example, gravity causes the Earth and the other planets to orbit the Sun, it also causes the Moon to orbit the Earth, and causes the formation of tides, the formation and evolution of the Solar System, stars and galaxies. The earliest instance of gravity in the Universe, possibly in the form of quantum gravity, supergravity or a gravitational singularity, along with ordinary space and time, developed during the Planck epoch, possibly from a primeval state, such as a false vacuum, quantum vacuum or virtual particle, in a currently unknown manner. Attempts to develop a theory of gravity consistent with quantum mechanics, a quantum gravity theory, which would allow gravity to be united in a common mathematical framework (a theory of everything) with the other three forces of physics, are a current area of research.

In 1687, English mathematician Sir Isaac Newton published Principia, which hypothesizes the inverse-square law of universal gravitation. In his own words,

"Ideduced that the forces which keep the planets in their orbs must [be] reciprocally as the squares of their distances from the centers about which they revolve: and thereby compared the force requisite tomkeep the Moon in her Orb with the force of gravity at the surface of the Earth; and found them answer pretty nearly."

Newton's theory enjoyed its greatest success when it was used to predict the existence of Neptune based on motions of Uranus that could not be accounted for by the actions of the other planets. Calculations by both John Couch Adams and Urbain Le Verrier predicted the general position of the planet, and Le Verrier's calculations are what led Johann Gottfried Galle to the discovery of Neptune.

A discrepancy in Mercury's orbit pointed out flaws in Newton's theory. By the end of the 19th century, it was known that its orbit showed slight perturbations that could not be accounted for entirely under Newton's theory, but all searches for another perturbing body (such as a planet orbiting the Sun even closer than Mercury) had been fruitless. The issue was resolved in 1915 by Albert Einstein's new theory of general relativity, which accounted for the small discrepancy in Mercury's orbit. This discrepancy was the advance in the perihelion of Mercury of 42.98 arcseconds per century.

Although Newton's theory has been superseded by Einstein's general relativity, most modern non-relativistic gravitational calculations are still made using Newton's theory because it is simpler to work with and it gives sufficiently accurate results for most applications involving sufficiently small masses, speeds and energies.

In general relativity, the effects of gravitation are ascribed to spacetime curvature instead of a force. The starting point for general relativity is the equivalence principle, which equates free fall with inertial motion and describes freefalling inertial objects as being accelerated relative to non-inertial observers on the ground. In Newtonian physics, however, no such acceleration can occur unless at least one of the objects is being operated on by a force.

Einstein's theory has important astrophysical implications. For example, it implies the existence of black holes— regions of space in which space and time are distorted in such a way that nothing, not even light, can escape—as an endstate for massive stars. There is ample evidence that the intense radiation emitted by certain kinds of astronomical objects is due to black holes. For example, microquasars and active galactic nuclei result from the presence of stellar black holes and supermassive black holes, respectively. The bending of light by gravity can lead to the phenomenon of gravitational lensing, in which multiple images of the same distant astronomical object are visible in the sky. General relativity also predicts the existence of gravitational waves, which have since been observed directly by the physics collaboration LIGO. In addition, general relativity is the basis of current cosmological models of a consistently expanding universe.

The Einstein field equations are nonlinear and very difficult to solve. Einstein used approximation methods in working out initial predictions of the theory. But as early as 1916, the astrophysicist Karl Schwarzschild found the first non-trivial exact solution to the Einstein field equations, the Schwarzschild metric. This solution laid the groundwork for the description of the final stages of gravitational collapse, and the objects known today as black holes. In the same year, the first steps towards generalizing Schwarzschild's solution to electrically charged objects were taken, which eventually resulted in the Reissner– Nordström solution, now associated with electrically charged black holes. In 1917, Einstein applied his theory to the universe as a whole, initiating the field of relativistic cosmology. In line with contemporary thinking, he assumed a static universe, adding a new parameter to his original field equations—the cosmological constant—to match that observational presumption. By 1929, however, the work of Hubble and others had shown that our universe is expanding. This is readily described by the expanding cosmological solutions found by Friedmann in 1922, which do not require a cosmological constant. Lemaître used these solutions to formulate the earliest version of the Big Bang models, in which our universe has evolved from an extremely hot and dense earlier state. Einstein later declared the cosmological constant the biggest blunder of his life.

General relativity has emerged as a highly successful model of gravitation and cosmology, which has so far passed many unambiguous observational and experimental tests. However, there are strong indications the theory is incomplete.The problem of quantum gravity and the question of the reality of spacetime singularities remain open. Observational data that is taken as evidence for dark energy and dark matter could indicate the need for new physics, even taken as is, general relativity is rich with possibilities for further exploration. Mathematical relativists seek to understand the nature of singularities and the fundamental properties of Einstein's equations, while numerical relativists run increasingly powerful computer simulations (such as those describing merging black holes).

3QUANTA

Quantum mechanics is the branch of physics relating to the very small. Quantization is one of the foundations of the much broader physics of quantum mechanics. Quantization of energy and its influence on how energy and matter interact (quantum electrodynamics) is part of the fundamental framework for understanding and describing nature.

In 1901, Max Planck used *quanta* to mean "quanta of matter and electricity", gas, and heat. In 1905, in response to Planck's work and the experimental work of Lenard, Albert Einstein suggested that radiation existed in spatially localized packets which he called "quanta of light"

The concept of quantization of radiation was discovered in 1900 by Max Planck, who had been trying to understand the emission of radiation from heated objects, known as black-body radiation. By assuming that energy can be absorbed or released only in tiny, differential, discrete packets, Planck accounted for certain objects changing colour when heated.On December 14, 1900, Planck reported his findings to the German Physical Society, and introduced the idea of quantization for the first time as a part of his research on black-body radiation. As a result of his experiments, Planck deduced the numerical value of h, known as the Planck constant, and reported more precise values for the unit of electrical charge and the Avogadro–Loschmidt number, the number of real molecules in a mole, to the German Physical Society.

One of the implications of quantum mechanics is that certain aspects and properties of the universe are **quantized**, i.e. they are composed of discrete, indivisible packets or **quanta**. For instance, the electrons orbiting an atom are found in specific fixed orbits and do not slide nearer or further from the nucleus as their energy levels change, but jump from one discrete quantum state to another. Even light, which we know to be a type of electromagnetic radiation which moves in waves, is also composed of quanta or particles of light called **photons**, so that light has aspects of both waves AND particles, and sometimes it behaves like a wave and sometimes it behaved like a particle (wave-particle duality).It results in what may appear to be some very strange conclusions about the physical world. At the scale of atoms and electrons, many of the equations of classical mechanics, which describe how things move at everyday sizes and speeds, cease to be useful. In classical mechanics, objects exist in a specific place at a specific time. However, in quantum mechanics, objects instead exist in a haze of probability; they have a certain chance of being at point A, another chance of being at point B and so on.

The principles of quantization, wave-particle duality and the uncertainty principle ushered in a new era for quamtum mechanics. In 1927, Paul Dirac applied a quantum understanding of electric and magnetic fields to give rise to the study of "quantum field theory" (QFT), which treated particles (such as photons and electrons) as excited states of an underlying physical field. Work in QFT continued for a decade until scientists hit a roadblock: Many equations in QFT stopped making physical sense because they produced results of infinity. After a decade of stagnation, Hans Bethe made a breakthrough in 1947 using a technique called "renormalization." Here, Bethe realized that all infinite results related to two phenomena (specifically "electron self-energy" and "vacuum polarization") such that the observed values of electron mass and electron charge could be used to make all the infinities disappear.

Since the breakthrough of renormalization, QFT has served as the foundation for developing quantum theories about the four fundamental forces of nature: 1) electromagnetism, 2) the weak nuclear force, 3) the strong nuclear force and 4) gravity. The first insight provided by QFT was a quantum description of electromagnetism through "quantum electrodynamics" (QED), which made strides in the late 1940s and early 1950s. Next was a quantum description of the weak nuclear force, which was unified with electromagnetism to build "electroweak theory" (EWT) throughout the 1960s. Finally came a quantum treatment of the strong nuclear force using "quantum chromodynamics" (QCD) in the 1960s and 1970s. The theories of QED, EWT and QCD together form the basis of the standard model of particle physics. Unfortunately, QFT has yet to produce a quantum theory of gravity. That quest continues today in the studies of string theory and loop quantum gravity.

4 A GENERAL FIELD THEORY

Some physicists believe that gravity is not a real force like the electromagnetic or the strong force because an inertial frame of reference eliminates the effects of gravity. One way to prove that gravity is a true force is if a gravitational field can be shown to possess energy density. When two masses undergo unsymmetrical acceleration, they emit quadrupole gravitational waves that definetely possess energy. A similar acceleration of two of the same polarity charged particles produces quadrapole electromagnetic radiation. The question is, if the electric and magnetic field store energy, what about the gravitational field? The first insight into answering the question came from Albert Einstein in his own words he stated that,

"to the intra-atomic movement of electrons, atoms would have to radiate not only electromagnetic but also gravitational energy if only in tiny amounts, as this is hardly true in nature, it appears that quantum theory would have to modify not only Maxwellian electrodynamics, but also the new theory of gravitation".

The problem of finding the energy density stored in the gravitational field reflects the incompleteness of general relativity. The problem of the free falling clocks which have to maintain the same rate is something related to the problem of non accountable gravitational energy. Feyman tried to put a patch on the issue of the gravitational energy but wasn't very successful either.

The problem is of paramount importance since it is reasonably something which impairs the unification of quantum theory and gravity. Newtonian gravitation is conservative globally and locally, GRT is not, quantum theory is based strictly on energy and momentum conservation.

A precise and consistent quantum theory of gravity has not yet been proved, not even by the self proclaimed geniuses of this time. We are aware and satisfied that classical General Relativity is the most precise description of gravity due to its predictable nature. The left hand side of Einstein field equation represents the metric of space time curvature while the right hand side represents the matter - energy content of the classical matter fields of pressure and energy density. It is known that quantum mechanics plays an important role in the behaviour of the matter fields but has no place in the Einsteins field equations.

In its simplest form the Einstein Field equation relates the cosmological constant Λ to the energy density ρ as,

$$\Lambda = \frac{8\pi G}{c^4}\rho$$

Where, G is the gravitational constant, and c is the constant speed of light.

The problem of reconciling the quantum theory with general relativity is brought about by not knowing the quantum mechanical version of the energy density, we only know the classical energy density in the electromagnetic field but the quantum mechanical energy density from which the electromagnetic energy density can be derived is not known. This is a very big problem faced by researchers in the field of quantum gravity. This wasn't only a problem to A. Einstein but also to Stephen.W.Hawking. In his paper "Particle creation by Black holes" Hawking wrote about the problem in this way, "one therefore has a problem of defining a consistent scheme in which the space time metric is treated classically but is coupled to the matter fields which are treated quantum mechanically"

By creating a correct quantum mechanical energy density from which both the electromagnetic and gravitational energy density can be derived we will be able to solve the quantum gravity problem. But before we do, let me present to you a quick insight into the Larmor power law which led to all of the problems we are facing today. Failure of the Bohr model and all quantum mechanics models to give a correct solution to the larmor problem is what is limiting us from finding the quantum theory of gravity.

For example, the laws of classical mechanics (i.e. the Larmor formula),

$$P = \frac{q^2 a^2}{6\pi\varepsilon_0 c^3}$$

Where a is the proper acceleration, q is the charge, and c is the speed of light.

predict that the electron will release electromagnetic radiation while orbiting a nucleus. Because the electron would lose energy, it would rapidly spiral inwards, collapsing into the nucleus on a timescale of around 16 picoseconds. This atom model is disastrous, because it predicts that all atoms are unstable. Also, as the electron spirals inward, the emission would rapidly increase in frequency as the orbit got smaller and faster. This would produce a continuous smear, in frequency, of electromagnetic radiation. However, late 19th century experiments with electric discharges have shown that atoms will only emit light (that is, electromagnetic radiation) at certain discrete frequencies. To overcome this hard difficulty, Niels Bohr proposed, in 1913, what is now called the *Bohr model of the atom*. He put forward these three postulates that sum up most of the model:

1. The electron is able to revolve in certain stable orbits around the nucleus without radiating any energy contrary to what classical electromagnetism suggests. These stable orbits are called stationary orbits and are attained at certain discrete distances from the nucleus. The electron cannot have any other orbit in between the discrete ones.

2. The stationary orbits are attained at distances for which the angular momentum of the revolving electron is an integral multiple of the reduced Planck's constant: , where $n = 1, 2, 3, ...$ is called the principal quantum number, and $\hbar = h/2\pi$. The lowest value of n is 1; this gives a smallest possible orbital radius of 0.0529 nm known as the Bohr radius. Once an electron is in this lowest orbit, it can get no closer to the proton. Starting from the angular momentum quantum rule, Bohr was able to calculate the energies of the allowed orbits of the hydrogen atom and other hydrogenlike atoms and ions. These orbits are associated with definite energies and are also called energy shells or energy levels. In these orbits, the electron's acceleration does not result in radiation and energy loss.The Bohr model of an atom was based upon Planck's quantum theory of radiation.

3. Electrons can only gain and lose energy by jumping from one allowed orbit to another, absorbing or emitting electromagnetic radiation with a frequency ν determined by the energy difference of the levels.

It must be carefully noted that the radiation from a charged particle carries energy and momentum. In order to satisfy energy and momentum conservation, the charged particle must experience a recoil force at the time of emission. The radiation must exert an additional force on the charged particle. This force is known as the Abraham– Lorentz force in the nonrelativistic limit and the Abraham–Lorentz–Dirac force in the relativistic setting. The Abraham–Lorentz force is the result of the most fundamental calculation of the effect of self-generated fields. It arises from the observation that accelerating charges emit radiation. The Abraham–Lorentz force is the average force that an accelerating charged particle feels in the recoil from the emission of radiation. The introduction of quantum effects leads one to quantum electrodynamics. The self fields in quantum electrodynamics generate a finite number of infinities in the view, however, the difficulties remain. Calculations can be removed by the process of renormalization. This has led to a theory that is able to make the most accurate predictions that humans have made to date.The renormalization process fails, however, when applied to the gravitational force. The infinities in that case are infinite in number, which causes the failure of renormalization. Therefore, general relativity has an unsolved self-field problem. String theory and loop quantum gravity are current attempts to resolve this problem, formally called the problem of radiation reaction or the problem of self force.

To overcome all the problems arising from the Larmor power formula we developed a new formula which relates the energy density to the force (and which is a generalized field formula) it is called the quantum mechanics energy density,

$$\rho = \frac{F^2}{8\pi\alpha hc}$$

Where ρ is the energy density stored in the field of the force F, α is the coupling constant that determines the strength of the force, and $\hbar$ is the reduced Planck constant.

Therefore the larmor power formula that was given above can be modified to be,

$$P = \frac{AF^2}{8\pi\alpha\hbar} = \frac{Am^2a^2}{8\pi\alpha\hbar}$$

Where, A is the surface area of orbit of a particle emmiting the radiations around the nucleus of an atom, m is the mass of the particle, and a is the proper acceleration.

The above given power formula will reduce to the Larmor power formula only when the area A is limited by,

$$A = \frac{4}{3\varepsilon_0 c^3}\left(\frac{q}{m}\right)^2 \alpha\hbar$$

Which implies that the area swept out by an electron in orbit around the nucleus of an atom is quantized. This is a classical version of loop quantum gravity in which the area of space occupied by a particle is quantized.

From the new larmor formula given we can clearly see that the charge does not appear in the formula. This then takes us to our original problem of determining the energy density in the gravitational field.

We know that the electric field store energy, and that in a vacuum the energy density is given by, $\rho = \frac{\varepsilon_0}{2}E^2$ where E is the electric Field and ε_0 the permittivity of free space. If our new formula for the enegy density given above is true, it must be able to reproduce the expression for the energy density of the electric field and also solve other problems.

To derive the energy density in the electric field, we let the force on the particle say an electron with charge e due to the electric field E created by another charged electron be, F=eE. Then the energy density will be related to the electric field by,

$$\rho = \frac{e^2 E^2}{8\pi\alpha hc}$$

But because the coupling constant of the electromagnetic force is the fine structure constant $\alpha = \frac{e^2}{4\pi\varepsilon_0 hc}$, then on substitution and cancelling like terms, we recover the energy density in the electric field as,

$$\rho = \frac{\varepsilon_0}{2} E^2$$

Similary, for the energy density in the gravitational field, let the force experienced by a particle of mass m due to the gravitational field g be F=mg. The energy density is here given by,

$$\rho = \frac{m^2 g^2}{8\pi\alpha hc}$$

But because the coupling constant of the gravitational force is the fine structure constant- $\alpha = \frac{Gm^2}{hc}$, then on substitution and cancelling like terms, we recover the energy density in the gravitational field as,

$$\rho = \frac{g^2}{8\pi G}$$

We have shown that, just as the electromagnetic field stores energy, the same is also true for the gravitational field.

Then the Einstein field equation can be summarized in the following format,

$$\Lambda = \frac{8\pi G}{c^4} \rho = \frac{8\pi G}{c^4} \left(\frac{F^2}{8\pi\alpha hc} \right)$$

$$\Lambda = \frac{GF^2}{\alpha hc^5} = \frac{F^2}{\alpha E_{pl}{}^2}$$

Where $E_{pl} = \sqrt{\frac{hc^5}{G}}$ is the Planck energy in Planck units.

We have therefore proved that the space time metric which is treated classically can be coupled to the matter fields which are treated quantum mechanically by the introduction of the energy density that is treated quantum mechanically.

Therefore in a limit where $F = \frac{c^4}{G}$ (Newtonian limit), we have the cosmological constant given by,

$$\Lambda = \frac{c^3}{\alpha hG} = \frac{1}{\alpha l_p{}^2}$$

Where $l_p = \sqrt{\frac{hG}{c^3}}$ is the Planck length.

We have shown that the unification of quantum mechanics with general relativity implies that there is a fundamental length in Nature in the sense that no operational procedure would be able to measure distances shorter than the Planck length. Finally, using hand waving

arguments we have also shown that a minimal length might be related to the cosmological constant which, if this scenario is realized, is time dependent.

The coupling constant is now related to the force between particles and to the cosmological constant by the following formula,

$$\alpha = \frac{GF^2}{\Lambda hc^5} = \frac{F^2}{\Lambda E_{pl}{}^2}$$

Such that when the force between two particles is the gravitational force $F = \dfrac{GMr}{R^2}$, and $\Lambda = \dfrac{R_s^2}{4R^4}$, ($R_s$ is Schwarzchild radius) we get the usual known gravitational coupling constant, $\alpha = \dfrac{GMm}{\hbar c} = \dfrac{E_g^2}{E_{pl}^2}$ (E_g is Schwarzchild Energy). This gives a simplest way of calculating the coupling constant.

Application of the quantum energy density to singularity resolution, information paradox, Planck stars, Emergence of the laws of Newton, galaxy rotation problem and the Tully-Fisher relationshi:-

We consider the possibility that the energy of a collapsing star and any additional energy falling into the Black hole could condense into a highly compressed core with density of the order of the Planck density. If we let the quantum force pressing on the surface of a star be given as,

$$F_q = \frac{hc}{R^2}$$

Where R is the radius of a Black hole, ℏ is the reduced Planck constant and c is the constant speed of light.

Let also the gravitational attraction force opposing the quantum force from within the collapsing star be

$$F_b = \frac{c^4}{G\alpha}$$

Where G is the gravitational constant and α is the coupling constant

The quantum energy density is given as,

$$\rho = \frac{F^2}{8\pi\alpha hc}$$

Subsitituting for F_b in the above given energy density we have,

$$\rho = \frac{c^7}{8\pi\alpha^3 hG^2}$$

Therefore nature appears to enter the quantum gravity regime when the energy density of matter reaches the Planck scale. The point is that this may happen well before relevant lengths become planckian. For instance, a collapsing spatially compact universe bounces back into an expanding one. The bounce is due to a quantum-gravitational repulsion which originates from the modified Heisenberg uncertainty, and is akin to the force that keeps an electron from falling into the nucleus. Therefore bounce does not happen when the universe is of planckian size, as was previously expected; it happens when the matter energy density reaches the Planck density. At this energy density, a Planck star is formed. The key feature of this theoretical object is that this repulsion arises from the energy density, not the Planck length, and starts taking effect far earlier than might be expected. This repulsive 'force' is strong enough to stop the collapse of the star well before a singularity is formed, and indeed, well before the Planck scale for distance. Since a Planck star is calculated to be considerably larger than the Planck scale for distance, this means there is adequate room for all the information captured inside of a black hole to be encoded in the star, thus avoiding information loss.

If this is the case, the gravitational collapse of a star does not lead to a singularity but to one additional phase in the life of a star: a quantum gravitational phase where the gravitational attraction is balanced by a quantum pressure and that is, when $F_q = F_b$

$$R = \alpha^{1/2}\left(\frac{hG}{c^3}\right)^{1/2}$$

$$R = \alpha^{1/2} l_p$$

Where $l_p = \left(\frac{hG}{c^3}\right)^{1/2}$ is the Planck length

For instance, if n = 1/3, and α is the gravitational coupling constant, a stellar-mass black hole would collapse to a Planck star with a size of the order of 10^{-10} centimeters. This is very small compared to the original star in fact, smaller than the atomic scale but it is still more than 30 orders of magnitude larger than the Planck length. This is the scale on which we are focusing here. The main hypothesis here is that a star so compressed would not satisfy the classical Einstein equations anymore, even if huge compared to the Planck scale. Because its energy density is already planckian.

Newton's law of gravity

Modified Newtonian dynamics (MOND) is a theory that proposes a modification of Newton's laws to account for observed properties of galaxies. It is an alternative to the theory of dark matter in terms of explaining why galaxies do not appear to obey the currently understood laws of physics. By applying our quantum mechanical energy density we explain why gravity is emergent and why the velocities of stars in galaxies were observed to be larger than expected based on Newtonian mechanics.

From the energy density relationship, the new effective gravitational force is related to quantum energy density by,

$$F = (8\pi\alpha hc\rho)^{1/2}$$

Since energy density is the amount of energy stored in a given system or region of space per unit volume, let a body of mass M store an amount of energy $E = Mc^2$ (due Einstein mass energy relationship) in a given region per unit volume $V = \frac{4}{3}\pi R^3$ where R is the radius. The energy density of this volume of space will be given by,

$$\rho = \frac{3Mc^2}{4\pi R^3}$$

If the dimensionless coupling number is given by,

$$\alpha = \alpha_g \frac{R_s}{12R}$$

Where $R_s = \frac{2GM}{c^2}$ is the Schwarzichild radius of a body of mass M , and $\alpha_g = \frac{Gm^2}{hc}$ is the gravitational coupling constant of a particle of mass m.

From the above given expression, the coupling number increases with an increase in the Schwarzchild radius but decreases with an increase in the radius R. For a particle m at a distance R so great from M the coupling number is very small.

When the expression of the energy density and coupling number are substituted into our force formula above, we get the usual Newton's law of gravitation as,

$$F = \left(8\pi\alpha_g \frac{R_s}{12R} hc \left(\frac{3Mc^2}{4\pi R^3} \right) \right)^{1/2}$$

$$F = \frac{GMm}{R^2}$$

From the above given derivation it has been shown that a particle m feels a force (i.e gravity) due to the energy density of mass M. The mass m comes from the gravitational coupling constant. This simply shows that gravity originates from the vacuum energy density and distance originates from the coupling number.

The Tully-Fisher Relation

While Newton's laws predict that stellar rotation velocities should decrease wih distance from the galactic centre, Rubin and collaborators found instead that they remain almost constant. The rotation curves are said to be flat. This observation necessitates either one of the following, 1) there exists in galaxies large quantities of unseen matter which boosts the stars velocities beyond what would be expected on the basis of the visible mass alone, or 2) Newton's laws do not apply to galaxies. The former leads to the dark matter hypothesis; the latter leads to Modified Newtonian dynamics (MOND)

Newton's laws works well in high acceleration environments, that is in the solar system and on Earth while it fails for objects with extremely low acceleration, such as stars in the outer parts of galaxies. To resolve the problem we have proposed a new effective gravitational force law given by,

$$F = (8\pi\alpha\hbar c\rho)^{1/2}$$

Applying this to an object of mass m in circular orbit around a point mass M (a crude approximation for a star in the outer regions of a galaxy), we find:

$$\frac{mv^2}{R_1} = (8\pi\alpha\hbar c\rho)^{1/2}$$

Since energy density is the amount of energy stored in a given system or region of space per unit volume, let a body of mass M store an amount of energy $E = Mc^2$ (due Einstein mass energy relationship) in a given region per unit volume V^*. The energy density of this volume of space will be given by,

$$\rho = \frac{Mc^2}{V^*}$$

Since the gravitational coupling constant for mass m is known to be,

$$\alpha = \frac{Gm^2}{\hbar c}$$

On substitution and cancelling like terms, we have

$$v^4 = \frac{8\pi c^2 R_1^2}{V^*}GM$$

Where v is the star's rotation velocity at a distance R_1 from the center of the galaxy, the rotation curve becomes flat, as required only when the following relation is correct,

$$\frac{V^{*}}{R_1{}^2} = 1.885 \times 10^{28}m = R_o$$

Which gives

$$v^4 = \frac{8\pi c^2}{R_o}GM = a_o GM$$

That is, the star's rotation velocity is independent of R_1, its distance from the centre of the galaxy- the rotation curve is flat, as required. By fitting this law to rotation curve data, we have found the Milgrom acceleration $a_o = 1.2 \times 10^{-10}m/s^2$.

For a sun's orbit around our galaxy, the radius of the sun's orbit is $R_1 = 8000pc$, $(1pc= 3.0857 \times 10^{16}m)$ this implies a volume of the space bound by matter at the centre of our milky way galaxy to be, $V^{*} = 1.149 \times 10^{69}m^3$.

This simple law is sufficient to make predictions for a broad range of galactic phenomena.

For other applications of the quantum mechanical energy density visit PartIII Section 13 on the Cosmological constant Problem.

5EVIDENCE FOR MAXIMAL ACCELERATION

Under the assumption of $\mu = m\alpha^{1/2}$ (where α is the coupling constant), in the Caianeillo maximum acceleration model ($A_{max} = \dfrac{\mu c^2}{m\lambda}$) , we derive the maximal acceleration and minimum radius to which a gravitating body can collapse in the commoving frame for both the Schwarzschild and Reissner-Nordstrom Black hole.

In the context of a geometrical unification of quantum mechanics and general relativity in phase space, Caianiello was the first person to propose the existence of a maximal proper acceleration for massive particles. Caianiello was able to derive the value $A_{max} = \dfrac{2mc^3}{\hbar}$ (1) for the maximum acceleration of a particle of rest mass m from the time-energy uncertainty relation. Caianiello model was based on two assumptions; $\hbar = \lambda \mu c$ and $\mu = 2m$ (2) for $A_{max} = \dfrac{\mu^2 c^3}{m\hbar} = \dfrac{c\hbar}{m\lambda^2} = \dfrac{\mu c^2}{m\lambda}$ (3).

Applications of Caianiello's model include cosmology, the dynamics of accelerated strings, neutrino, oscillations and the determination of a lower neutrino mass bound. There is also evidence for maximal acceleration and singularity resolution in covariant loop quantum gravity found by Rovelli and Vidotto.

In this book we propose an adhoc assumption of $\mu = m\alpha^{1/2}$ where α is the coupling constant. This differs from Caianiello's model assumption of $\mu = 2m$. Therefore the maximum acceleration(3) will be given by,

$$A_{max} = \frac{c^2}{r}\alpha^{1/2}$$

(1)

Where, r is the smallest possible distance between any two masses. In this book r takes values for the Schwarzschild and Reissner-Nordstrom radius.

Note: Equation (1) given above reduces to the value $A_{max} = \dfrac{2mc^3}{h}$ that was earlier derived by Caianiello under two conditions;

(i) When $r = \dfrac{2GM}{c^2}$ and $\alpha = 16\alpha_g^{\,2}$ for a Schwarzschild Black hole of mass M. Where $\alpha_g = \dfrac{GMm}{\hbar c}$ is the gravitational coupling constant

(ii) When $r = \left(\dfrac{Ge^2}{4\pi\varepsilon_0 c^4}\right)^{1/2}$ and $\alpha = 4\alpha_g\alpha_e$ for a Reissner-Nordstrom(RN) Black hole. Where $\alpha_e = \dfrac{e^2}{4\pi\varepsilon_0\hbar c}$ is the electromagnetic coupling constant and M=m.

We have therefore recovered the Caianiello maximum acceleration for the schwarzchild black hole (BH) and RN black hole. We have found that the change in the coupling constant depends on the type of the black hole radius while keeping the acceleration the same for both bodies falling into the hole. This therefore means that, the laws of physics will follow a different path for each black hole. In the schwarzichild black hole the coupling will follow a power law in the gravitational coupling constant while in the RN black hole the coupling is the product of the gravitational and electromagnetic coupling meaning that, the RN black hole will create an atmosphere of unified electromagnetic and gravitational fields.

In conclusion therefore, the acceleration of any celestial body will not be affected by the radius of the black hole provided the right choice of the coupling constant is choosen for the black hole in question.Therefore coupling is an profound property of a black hole. Let it be known from here that,

Whereas the Schwarzichild black hole has a radius $r = \dfrac{2GM}{c^2}$ there must be a coupling constant associated with it given by $\alpha = 16\alpha_g^{\,2}$, which is named the coupling constant for a Schwarzschild BH. Also Whereas the RN black hole has a radius $r = \left(\dfrac{Ge^2}{4\pi\varepsilon_0 c^4}\right)^{1/2}$ there must be a coupling constant associated with it given by $\alpha = 4\alpha_g\alpha_e$, which is named the coupling constant for a RN BH.

Note: the above statements are in accordance with the unification of gravity with quantum mechanics and should not be confused with classical general relativity as we are to see below. The coupling constants given are as a result of the derivation of the Caianiello acceleration and are bound to change in other derivations as we are to see below.

Maximal Acceleration in Quantum Gravity

(i)Reissner-Nordstrom black hole

Considering the event horizon of a Reissner-Nordstrom black hole of radius $r = \left(\dfrac{Ge^2}{4\pi\varepsilon_0 c^4}\right)^{1/2}$ and gravitational coupling $\alpha = \dfrac{GM^2}{\hbar c}$ on assumption that M=m. Then substituting in (1), the growing acceleration approaching a classical singularity in the Reissner-Nordstrom metric is bounded by the existence of a maximal acceleration of;

$$a_{max} = \frac{M}{e}\left(\frac{4\pi\varepsilon_0 c^7}{\hbar}\right)^{1/2}$$

(2)

Where e is charge on an electron, ε_0 is the permittivity of free space and $\hbar$ is the reduced Planck constant. The above formula shows that the acceleration of a body in the RN BH depends on the matter content and charge of the BH itself.

(ii)Schwarzschild Black hole

Considering the event horizon of a Schwarzschild black hole of radius $r = \dfrac{GM}{c^2}$ and gravitational coupling $\alpha = \dfrac{GM^2}{\hbar c}$. Then substituting in (1), the growing acceleration approaching a classical singularity in the Schwarzschild metric is bounded by the existence of a maximal acceleration of;

$$a_{max} = \left(\frac{c^7}{Gh}\right)^{1/2}$$

(3)

The above derivations show that the coupling constants expressions used in the derivation of the Caianiello acceleration are different from those used here. This implies that when moving from classical mechanics to quantum mechanics the coupling constant is $\alpha = 16\alpha_g^2$. And when moving from quantum mechanics to quantum gravity, the coupling constant is $\alpha = \alpha_g$ for a schwarzichild blackhole.

6 EVIDENCE FOR MINIMAL LENGTH

Because of the equivalence principle in the case of gravitational interaction, we propose to show here that the existence of a minimal length for both a Reissner and Schwarzschild Black hole is a straight forward consequence of our maximal acceleration value (1).

In Newtonian law (center of mass system)

$$\frac{GM}{R^2} = \frac{c^2}{r}\alpha^{1/2}$$

Where, R is the radius of a Black hole (In this case the minimum radius to which a central mass will collapse)

On arranging we have,

$$R = \frac{1}{\alpha^{1/4}}\left(\frac{R_s r}{2}\right)^{1/2}$$

$$(4)$$

Where R_s is the Schwarzschild radius $R_s = \frac{2GM}{c^2}$.

Two results are thus deduced;

i) For $r = \left(\frac{Ge^2}{4\pi\varepsilon_0 c^4}\right)^{1/2}$ the radius of the event horizon of a Reissner Black hole

and $\alpha = \frac{GM^2}{\hbar c}$, the minimum radius to which a gravitating body will collapse in a commoving frame of the Reissner-Nordstrom metric will have a value;

$$R_{min} = \left(\frac{\hbar e^2 G^2}{4\pi\varepsilon_o c^7}\right)^{1/4} = \alpha_e^{1/4} l_p$$

$$. (5)$$

Where l_p is the Planck length and α_e is the fine structure constant $1/137$.

$$R_{min} = 4.724 \times 10^{-36} m$$

This length is smaller than the Planck length and therefore imposes a naked singularity.

ii) Similarly, for $r = \frac{GM}{c^2} = \frac{R_s}{2}$ the radius of the event horizon of a Schwarzschild Black hole and $\alpha = \frac{GM^2}{\hbar c}$, the minimum radius to which a gravitating body will collapse in a commoving frame of the Schwarzschild metric will have a value;

$$R_{min} = (M)^{1/2}\left(\frac{\hbar G^3}{c^7}\right)^{1/4}$$

$$. (6)$$

The above derivation clearly provides evidence for the existence of a maximal acceleration and minimal length which are both expected in the theory of quantum gravity to cure strong singularities such as, big bang, big crunch, black holes etc.

We have clearly modified the geometry of Rindler space by the introduction of the coupling $\alpha^{1/2}$ into the formula for acceleration. We have witnessed that the presence of $\alpha^{1/2}$ into the formula for acceleration leads to an exact evidence for the existence of the maximal acceleration and minimal length for both the Reissner-Nordstrom and Schwarzschild black holes in quantum gravity. The split horizon in a Rindler wedge at a distance R= c^2/a for the acceleration a has been modified here, hope you have witnessed how $\alpha^{1/2}$ changes all of this. This implies that there is some fundamental limitation on how much acceleration a particle could experience based on the strong-field behavior of the fundamental force causing it.

Results of the maximal acceleration and minimal length for the Reissner Black hole have not been derived anywhere in literature. These clearly impose a general bound on acceleration and length (in Reissner space time geometry) with implications. For example, a black hole the size of an electron ($m_e = 9.11 \times 10^{-31} \text{kg}$), imposes an acceleration of $a_{max} = 2.732 \times 10^{30} \text{m/s}^2$. So this accelerated frame would detect a Unruh radiation at 1.1×10^{10} K. Also the minimal length result implies the existence of the discreteness (granular nature) of space and cures the singularities that plague General relativity by imposing a general bound on length of $4.724 \times 10^{-36} \text{m}$.

In conclusion, a corrected Rindler space geometry directly proves an existence for the maximal acceleration and minimal length in quantum gravity, not only for the Schwarzschild metric with a horizon distance half of the Schwarzschild radius but also for the Reissner metric. Therefore the introduction of $\alpha^{1/2}$ in the formula for acceleration must be thoughtfully investigated as this solves all the problems brought about by the General relativity theory.

7EVIDENCE FOR MINIMAL BLACK HOLE MASS

The smallest black hole would be one where the Schwarzschild radius equals the radius of a mass with a reduced Compton wavelength which is the smallest size to which a given mass can be localized. For a small mass M, the Compton wavelength exceeds half the Schwarzschild radius, and no black hole description exists. This smallest mass for a black hole is thus approximately the Planck mass, the micro black hole.

Contrary to the above observation, torsion (see Einstein-Cartan theory) modifies the Dirac equation in the presence of the gravitational field causing fermions to be spatially extended. This spatial extension of fermions limits the minimum mass of a black hole to be on the order of $10^{16} Kg$, showing that micro black holes (of Planck mass) may not exist. Another mass limit is from the data of the Fermi Gamma-ray space telescope satellite which states that, less than one percent of dark matter could be made of primordial black holes with masses up to $10^{13} Kg$.

The major aim of this section is to prove theoretically the existence of a minimum mass limit of a black hole and thereafter prove Chandrasekhar wrong (see Chandrasekhar 1983 Noble lecture concluding statement below)

"We conclude that there can be no surprises in the evolution of stars of mass less than 0.43Solarmass ($\mu = 2$). The end stage in the evolution of such stars can only be that of the white dwarfs. (**Parenthetically, we may note here that the so-called 'mini' black-holes of mass $10^{12} Kg$ cannot naturally be formed in the present astronomical universe.**)"

From the theory of white dwarf stars, the radius limit of a white dwarf of mass M is given by the following equation,

$$R_W = \frac{(9\pi)^{2/3}\,h^2}{8} \frac{1}{m_e G (m_{pro})^{5/3} M^{1/3}}$$

(i)

Where m_{pro} and m_e is the proton and electron mass respectively

Just like the Compton wavelength, there must exist another radius for the consititution of stars that differs from the radius given in (i) above. For example, in the same way the Planck mass is deduced (i.e by equating the Schwarzschild radius to the Compton wavelength) is the same way in which we are to prove the existence of the mass limit of a black hole.

We start from first principles. Let it be known that the derivation of the Chandrasekhar mass limit will follow the equipartition of the gravitational potential energy of a star to its electron degeneracy pressure. Where by, if the gravitational binding energy is given by,

$$E_g = \frac{2 M_{pl}{}^3 m_e c^2 (6.144\pi^3)}{M m_{pro}{}^2 \quad \mu^2}$$

Where M_{pl}, is the Planck mass and μ is the average molecular weight per electron

And the electron degeneracy energy pressure of the star is given by,

$$E_d = m_e c^2$$

When $E_g = E_d$ then we obtain the mass limit of the white dwarf star as,

$$M = \frac{12.288\pi^3 M_{pl}{}^3}{\mu_e{}^2 \quad m_p{}^2} = 1.4 M_{sun}$$

If then this is true, then the formula for the gravitational binding energy of a star is also true. This there fore implies that the following assumption will also be true.

When the binding gravitational energy of a star is equal to the Newtonian gravitational potential energy $\frac{GM^2}{R}$ we obtain the radius which is the smallest size to which a given mass of a star can be localized as,

$$\frac{GM^2}{R} = \frac{2M_{pl}{}^3 m_e c^2 (6.144\pi^3)}{Mm_{pro}{}^2 \quad \mu^2}$$

$$R = \frac{Gm_{pro}{}^2}{2m_e c^2}\left(M\big/M_{pl}\right)^3 \frac{\mu^2}{6.144\,\pi^3}$$

(ii)

This can be rewritten in the form,

$$R = R_k \left(M\big/M_{pl}\right)^3$$

Where $R_k = 2.384 \times 10^{-53}m$ which is smaller than the Planck length of $1.62 \times 10^{-35}m$

Therefore equating Equation (i) to Equation (ii) we deduce the mass limit of a black hole as,

$$M = \left(293.534\pi^{11}M_{pl}^{21} \middle/ \mu^6 M_{pro}^{11} \right)^{1/10} = 9.54 \times 10^{13} Kg$$

The value is in excellent agreement with other theoretical and experimental observations

The radius of this black hole from Equation (ii) is thus $2.527 \times 10^{14}m$ larger than the radius of the sun of $7 \times 10^8 m$.

In conclusion therefore the end stage in the evolution of a star can only be that of the black hole with a mass $9.54 \times 10^{13} Kg$ and size of $2.53 \times 10^{14}m$ in contrast with the Chandrasekhar observations.

Note that the radius given by Equation (ii), $R = R_k \left(M / M_{pl} \right)^3$ above is similar to the Equation for the size of the Planck star that was given by Rovelli and Vidotto, $r = l_p \left(\frac{M}{M_{pl}} \right)^n$ where l_p is the Planck length and n is the positive number. This is a clear indication that there is a length that is smaller than the Planck length.

8 EVIDENCE FOR MINIMAL AND MAXIMAL MAGNETIC FIELD STRENGTH

The scale in quantum electrodynamics (QED), above which the electromagnetic field is expected to become non linear, also called the Schwinger limit, was first derived by Fritz Sauter in 1931. However In this section we develop a mechanism (which differs from Fritz's approach) through which the Schwinger limit is deduced using a dimensionless number, which gives the critical magnetic field in quantum electrodynamics when its value is equal to the electromagnetic coupling constant and in the same way gives the critical magnetic field in Quantum gravity when its value is equal to the gravitational coupling constant.

According to D.A. Leahy, the application of quantum electrodynamics in strong magnetic fields only fairly recently has been a subject of interest. The motivation for this study was the discovery of Neutron stars with very high magnetic fields of orders 10^{12} -10^{13}G.

With the discovery of magnetars, quantum electrodynamics calculations which are valid for very high fields became of great interest. The critical value of the magnetic field is defined as $B = \dfrac{m^2 c^2}{\hbar e} = 4.414 \times 10^{13} G$ .However, there is a value of the magnetic field that is bigger and stronger than the critical magnetic field strength in Quantum electrodynamics and this magnetic field is of orders of magnitude $10^{52} G$. Such a big value has not been deduced in any existing scientific literature and that is the reason why I take pleasure in deriving it here and hence call it the "quantum gravity threshold".

From the intensity wave equation, $\dfrac{F_B c}{A} = \dfrac{F_e^2}{2nh}$, if we let the magnetic force to be equal in magnitude and strength to the electric force, we create two relationships, 1) the force on a particle falls off as the area it occupies and 2) the force falls off as the principle quantum number.

$$\text{Force (F)} = \frac{2hc}{A} = \frac{Bev}{n}$$

If n was the fine structure constant ($ke^2/\hbar c$, $k = 1/4\pi\varepsilon_0$), the speed of light in vacuum being $c = \lambda f = \lambda\omega/2\pi$ and the velocity of a particle in the magnetic field is $v = \omega r$ where ω is the angular frequency for circular motion we have

$$\frac{F_c}{F} = \frac{em}{2B\lambda\hbar\varepsilon_0}$$

Where $F_c = m\omega^2 r$ is the centripetal force

We have thus derived a general formula for the coupling of forces. Then the Schwinger limit in quantum electrodynamics for the critical magnetic field can be deduced from the above expression when we set the ratio of the forces to be equal to the electromagnetic coupling or fine structure constant as, $\frac{F_c}{F} = \frac{ke^2}{hc}$

$$B_{QED} = \frac{2\pi mc}{\lambda e}$$

For a particle with deBrogile wavelength $\frac{2\pi\hbar}{mc} = \lambda$, the quantum electrodynamics threshold is given by,

$$B_{QED} = \frac{m^2 c^2}{\hbar e} = 4.3697 \times 10^{13} G$$

However, for $\frac{F_c}{F} = \frac{Gm^2}{hc}$, the gravitational coupling constant, and $\frac{2\pi\hbar}{mc} = \lambda$, the deBrogile wavelength, the quantum gravity threshold is given by a value,

$$B_{QG} = \frac{ec^2}{4\pi G\hbar\varepsilon_o} = 1.8423 \times 10^{52}\,G$$

We have thus deduced the constant magnetic field carried by an electron in the combined quantum electromagnetic and gravitational fields. The fact that the formula has the fundamental constant of electricity (ε_o), relativistic quantum mechanics (c, $\hbar$) and Gravity (G), is an indication that this is the quantum gravity limit or a scale at which the electromagnetic field is expected to become non linear.

In conclusion, the dimensionless coupling constant is therefore the ratio of the universal constant magnetic field (B_0) on the particle of mass m to the Schwinger magnetic induction limit- the strong magnetic field B external to virtual electron-positron pair enclosed in a quantum vacuum. In general, the coupling constant can be written as,

$$\alpha = \frac{em}{2B\lambda\hbar\varepsilon_o}$$

For $\lambda = \frac{2\pi\hbar}{mc}$, the coupling constant reads as, $\alpha = \frac{em^2c}{4\pi B\lambda\hbar^2\varepsilon_o}$

Implying, $\alpha = \frac{B_o}{B}$

Where $B_o = \left(\frac{m}{\hbar}\right)^2 \frac{ec}{4\pi\varepsilon_o} = 3.22 \times 10^{11}\,G$,

This value of the magnetic field is smaller than the Schwinger limit and therefore has implications for the theory of QED. Once such a value is discovered in experiments it will lead to a complete observation of the electron-positron pairs. Finally the fine structure coupling constant and the gravitational coupling will manifest in situations where the magnetic field B is either a Schwinger limit given by equations above. This implies that, the value of the magnetic field B_0 derived above is the same everywhere (i.e both in applications of electromagnetism and gravity)

PART II ELABORATIONS

9 THERE'S NO SUCH THING AS THE FORCE OF GRAVITY

Is Gravity and the Laws of Physics Emergent?

Starting from first principles and general assumptions we present a heuristic argument that shows that Newton's law of gravitation and Coloumb's law of electricity naturally arise in a theory in which space emerges through a zero- point fluctuation of the quantum vacuum. Gravity is identified with a casimir force caused by quantum vacuum fluctuations due to the presence of material bodies in it or the distortion of the vacuum through its interaction with mass. A relativistic generalization of the presented arguments directly leads to the Einstein equations. When space is emergent even Newton's law of inertia needs to be explained. The equivalence principle suggests that it is actually the law of inertia whose origin is casimir.

The real origin of gravity is one of the most important, complex and substantially yet unsolved questions in Physics. The replacement of the Newtonian model of gravity with the Einstein's one given by General Relativity (GR) has only shifted the question without solving it. Within GR, gravity has two possible interpretations: a field one and a geometric one. According to the latter, that has become the prevalent one, gravity is due to the curvature of the space – time "tissue", represented as a "rubber sheet", due to the presence of a mass. Nevertheless, this is a purely mathematical description telling nothing about the physical mechanism starting the motion. In fact, even supposing the existence, in the neighbouring of a source mass, of a curved four – dimensional manifold it doesn't explain why a second particle at rest should move towards the source mass.

As such, it invites attempts at derivation from a more fundamental set of underlying assumptions, and six such attempts are outlined in the standard reference book Gravitation, by Misner, Thorne, and Wheeler (MTW). ' Of the six approaches presented in MTW, perhaps the most far-reaching in its implications for an underlying model is one due to Sakharov; namely, *that gravitation is not a fundamental interaction at all, but rather an induced effect brought about by changes in the quantum fluctuation energy of the vacuum when matter is present.* ' In this view the attractive gravitational force is more akin to the induced van der Waals and Casimir forces, than to the fundamental Coulomb force. Although speculative when first introduced by Sakharov in 1967, this hypothesis has led to a rich and ongoing literature on quantum-fluctuation-induced gravity that continues to be of interest. In this approach the presence of matter in the vacuum is taken to constitute a kind of set of boundaries as in a generalized Casimir effect, and the question of how quantum fluctuations of the vacuum under these circumstances can lead to an action and metric that reproduce Einstein gravity has been addressed from several viewpoints.

Therefore in this chapter we want to show that gravitation might be not a fundamental interaction but a byproduct of the electromagnetic interaction, precisely an electromagnetic phenomena induced by the presence of matter in the quantum vacuum (the quantum field that is present even in empty space). Which means that, matter is not just there but is in the quantum vacuum, and therefore interacts with it, causing some kind of quantum fluctuation energy, that fluctuation is gravitation. In simple terms, a body immersed in quantum fields will interact with them causing gravity to manifest.

(a)Emergence of the laws of Newton

Haisch, Rueda, and others have made the claim that the origin of inertial reaction forces can be explained as the interaction of electrically charged elementary particles with the vacuum electromagnetic zero-point field expected on the basis of quantum field theory.

Gravity is treated as a residuum force in the manner of casimir or vander waals forces. Expressed in the most rudimentary way this can be viewed as follows. The zero point field causes a given charged particle to oscillate. Such oscillations give rise to secondary electromagnetic fields. An adjacent charged particle will thus experience both the zero point field driving forces causing it to oscillate, and in addition forces due to the secondary fields produced by the zero point field driven oscillations of the first particle. Similarly, the zero point field driven oscillations of the second particle will cause their own secondary fields acting back upon the first particle. The net effect is an attractive force between the particles.

Force and Inertia

For the interaction between two particles, each mass experiences a background zero point field and a zero point driven dipole field of the other mass.

Two masses A and D (taken here to be equal for ease of discussion) with D located a distance R from A, along the positive z axis of a coordinate system centered at A. The zero point field will cause a charged particle A to oscillate. The oscillations will then give rise to a secondary electromagnetic field , which will cause particle D to oscillate. In the same way, the zero point field driven oscillations of particle D will cause their own secondary fields acting back upon particle A. the net effect will be an attractive force between particles A and D that will cause one to move towards the other with a small acceleration a_0 in the weak field limit.

Analogous to the Compton Effect, the wavelength of the electromagnetic waves emitted or scattered as a result of particle A interacting with the quantum vacuum will be given as,

$$\lambda_1 = \frac{2\pi R^2}{\lambda_c}\left(\frac{B_0}{B}\right)\alpha$$

Where,

B_0- is the strong magnetic field (greater than or equal to the critical value, which can create electron-positron pairs from the quantum vacuum). The Schwinger mechanism has two cornerstones, the first one is the existence of quantum vacuum and the second one the existence of an external electric field (which attempts to separate electrons and positrons). There are no particles in the vacuum (in that sense the vacuum is empty); but the vacuum is plenty of short-living virtual particle-antiparticle pairs which in permanence appear and disappear (allowed by time- energy uncertainty relation). A "virtual" pair can be converted into a real electron-positron pair only in the presence of a strong external field, which can spatially separate electrons and positrons, by pushing them in opposite directions, as it does an electric field. Therefore the zero point field or quantum vacuum exists but with an external magnetic field stronger than the critical value such that when a particle A is immersed in this zero point field or quantum vacuum, it will interact with the quantum vacuum causing quantum vacuum fluctuations which will trigger the external magnetic field causing oscillation of particle A giving rise to secondary electromagnetic fields.

B- is the value of the magnetic field that exists between particle A and D. This value depends only on the masses of the particles. In other words it is the magnetic field that depends on the matter constituents of the particles in question irrespective of the distance.

$$B = \frac{m^2 ec}{4\pi\varepsilon_o h^2}$$

λ_{c^-} is the reduced Compton wavelength $\frac{\hbar}{mc}$ and α is a dimensionless coupling constant.

It must be noted that, in the weak field limit, the resistance which defines the inertia of a particle is, ultimately, electromagnetic resistance caused by the zero point field on the particle, and it is this resistance which produces gravitational waves with a wavelength of due to a state of motion of a particle,

$$\lambda_2 = \frac{2\pi c^2}{a_o}\alpha$$

From the above given assumption, it is proposed that a body's inertia is due, to the distribution of matter in the universe, and, more precisely, to the electromagnetic interaction that arises from quantum fluctuations of the zero point field in accelerated frames. Basically, a particle's inertia is a function of the particle's interaction with zero point field. Inertia is resistance to acceleration and this reistance causes a form of the gravitational wave simply because resistance becomes a force. This implies that, the resistance which defines the inertia of a particle is, ultimately, electromagnetic resistance caused by the zero point field on the particle.

It must therefore be true that under a condition where, $\lambda_1 = \lambda_2$, we recover Newton's law of inertia (F=ma) as,

$$F = ma_o = \frac{\hbar c}{R^2}\left(\frac{B}{B_o}\right) \qquad (1)$$

Therefore matter continuosly interacts with the zero point field (as Casimir effect), and this interaction yields a force (the resistance to motion) whenever acceleration takes place. Inertia is due to the distortion of the zero point fluctuations in an accelerated reference frame. Technically, inertia is due to the high frequencies of the distortion of the zero point spectrum.

Newton's law of gravity

For the interaction between two masses, each mass experiences a background zero point field and a zero point field driven dipole field of the other mass. The procedure followed here is precisely that developed by Boyer for the derivation of the retarded van der waals forces at all distances between a pair of polarizable particles. Therefore we need only outline the procedure as it applies to our case.

Two masses A and D (taken here to be equal for ease of discussion) with D located a distance R from A, along the positive z axis of a coordinate system centered at A. The modified casimir force between the pair of particles A and D is given by Eqn1,

$$F = ma_o = \frac{\hbar c}{R^2}\left(\frac{B}{B_o}\right)$$

Where B_o is the external (or background) magnetic field stronger than the critical value and B is the dipole magnetic field at the position of particle A due to the motion of particle D and so forth. But since, $B_0 = \dfrac{ec^2}{4\pi\varepsilon_0 Gh}$, and $B = \dfrac{m^2 ec}{4\pi\varepsilon_0 h^2}$, on substitution into Eqn1 we obtain a familiar law,

$$F = \frac{Gm^2}{R^2}$$

We have recovered Newton's law of gravitation, practically from first principles!

These equations do not just come out by accident. It had to work, partly for dimensional reasons. In a sense we have reversed these arguments. But the logic is clearly different, and sheds new light on the origin of gravity: it is a casimir force! That is the main statement, which is new and has not been made before. If true, this should have profound consequences.

It is hereby proposed that, gravity is not a separately existing fundamental force, but rather a residuum force derived from zero-point fluctuations of other fields in the manner of the Casimir and van der Waals forces. Particularizing this hypothesis to the zero point fluctuation of the vacuum electromagnetic field, we identify the gravitational force as the casimir force associated with the long-range radiation fields (as opposed to the usual shorter-range induction fields) generated by the particle motion response to the zero point fluctuation of the electromagnetic field.

It is therefore seen that a well-defined, precise quantitative argument can be made that gravity is a form of long-range casimir force associated with particle response to the zero-point fluctuations of the electromagnetic field. As such, the gravitational interaction takes its place alongside the short-range van der Waals forces and the Casimir force as related phenomena which emerge from the underlying dynamics of the interaction of particles with the zero-point auctuations of the vacuum electromagnetic field.

(b)Emergence of electromagnetism

Electromagnetism or the coloumb force emerges in a similar fashion as the gravitational force. The origin of the electric force here assumes a critical magnetic field (Schwinger effect or limit) taken here to represent the external magnetic field,

$$B_0 = \frac{m^2 c^2}{\hbar e}$$

But since, $= \frac{m^2 e c}{4\pi\varepsilon_0 \hbar^2}$. On substitution into Eqn 1 we obtain a familiar law,

$$F = \frac{e^2}{4\pi\varepsilon_0 R^2}$$

We have recovered Coulomb's law of electricity, practically from first principles!

These equations do not just come out by accident. It had to work, partly for dimensional reasons. In a sense we have reversed these arguments. But the logic is clearly different, and sheds new light on the origin of electricity and gravity: it is a casimir force!

Therefore the gravitational field is the set of all electromagnetic fields generated by all particles as they interact with the zero point field. Gravity and electricity results from a distortion of the quantum vacuum through its interaction with a mass.

A more Simpler Derivation: Emergence of Gravity

The first attempt into the derivation of the gravitational force and Newton's laws of inertia was given in part by Erik Verlinde (2011) in which he stated that gravity is an entropic force. Simple as it was, his ideas were on a large scale rejected by mainstream physicists. The rejection of verlinde ideas where not backed up by another approach as it has been with other theories save for Sabine Hossenfelder approaches and critics. I think Erik was not surprised by these attacks because this is what physicists do especially when ones analysis or derivation doesn't involve the use of rigorous mathematical models-the one which were used by Einsten and others.

Anyway, what could be another approach towards the derivation of the gravitational, electricity and the law of inertia different from Verlindes idea of the emergent of gravity as an entropic force?

In this chapter we derive Newton's laws of inertia, gravitation and also the electromagnetic force law from first principles without assuming dark matter and the MOND theories. To differ from Verlindes approach we shall use the notions of Quantum vacuum and the Schwinger effect or limit in QED.

In order to understand the physical significance of the derivation to be given herein, we must remember the Schwinger mechanism (Schwinger, 1951) in Quantum Electrodynamics: a strong electric field, greater than a critical value, can create electron-positron pairs from the quantum vacuum.

The Schwinger mechanism has two cornerstones, the first one is the existence of quantum vacuum and the second one the existence of an external electric field (which attempts to separate electrons and positrons). There are no particles in the vacuum (in that sense the vacuum is empty); but the vacuum is plenty of short-living virtual particle-antiparticle pairs which in permanence appear and disappear (allowed by time- energy uncertainty relation). In simple words, the quantum vacuum is a kingdom of the virtual particle-antiparticle pairs; a kingdom with apparently perfect symmetry between virtual matter and virtual antimatter.

A "virtual" pair can be converted into a real electron-positron pair only in the presence of a strong external field, which can spatially separate electrons and positrons, by pushing them in opposite directions, as it does an electric field. Thus, "virtual" pairs are spatially separated and converted into real pairs by the expenditure of the external field energy. For this to become possible, the potential energy has to vary by an amount in the range of about one Compton wavelength , which leads to the conclusion that a significant pair creation occurs only in a very strong external field E, greater than the critical value.

Therefore the external force which attempts to separate particles and antiparticles converting a virtual pair into a real one may be simplified as,

$$F\Phi = 4\pi E_o h$$

(a)

Where F is the external force, Φ is the magnetic flux impeding a sphere of radius R and area $A = 4\pi R^2$ and E_o is the electric field on an electron of mass m

Derivation of force and inertia

The magnetic field is related to the electric field by

$$E_0 = B_0 c = \frac{Mmec^2}{4\pi\varepsilon_o h^2}$$

(b)

Remember the above given value is constant for a particle with mass m. There is also an assumption for M=m. Note also that $B_0 \neq B$

A new assumption and probably the most surprising one is that, the magnetic flux can be related to energy W and acceleration a by

$$\Phi = \frac{We}{ah\varepsilon_o}$$

(c)

Thus, "virtual" pairs are spatially separated and converted into real pairs by the expenditure of the external field energy. For this to become possible, the potential energy has to vary by an amount in the range of about one Compton wavelength , which leads to the conclusion that a significant pair creation occurs only in a very strong external field E, greater than the critical value E_o.

But, $W = Mc^2$, it is therefore evident why equation (c) for the magnetic flux was chosen to be of the given form. It is picked precisely in such a way that one recovers the second law of Newton

F=ma

As easily verified by combining (a) together with (b) and (c)

Therefore, a similarity or resemblance between acceleration, thermodynamics and electromagnetism comes alive in the statement below;

As there is a formula for the temperature T that is required to cause an acceleration equal to a, $T = \frac{ha}{2\pi kc}$ so there must also be a temperature required to cause an acceleration for an electron in the quantum vacuum at a constant electric and magnetic flux, $T = \frac{ha\varepsilon_o \Phi}{ke}$

Derivation of Newton's law of Gravity

Suppose our universe is a sphere of area $A = 4\pi R^2$ with a sea of virtual particles in a quantum vacuum. It is theorized as before that to separate the virtual particle and antiparticles in a vacuum into real particles one will require a strong external electric or magnetic field B. For the gravitational field, this external magnetic field was calculated to be,

$$B = \frac{ec^2}{4\pi G \hbar \varepsilon_o}$$

Where we introduced a new constant G. Eventually this constant is going to be identified with Newton's constant, of course. But since we have not assumed anything yet about the existence of a gravitational force, one can simply regard this equation as the definition of G.

Then, the magnetic flux will be given as,

$$\Phi = BA = \frac{ec^2 R^2}{G \hbar \varepsilon_o}$$

$$(\mathbf{d})$$

Substituting Equation (b) and (d) into (a) one obtains the familiar law

$$F = G\frac{Mm}{R^2}$$

We have recovered Newton's law of gravitation, practically form first principles. Following the above derivation carefully, it implies that gravity is a quantum force resulting from the quantum fluctuations of the vacuum due to an existence of an external strong electric or magnetic field separating particles from antiparticles or matter from anti matter.

Derivation of the law of electromagnetism

Following the same steps as in the previous derivation for the gravitational force but assuming a different external magnetic field (Schwinger limit),

$$B = \frac{Mmc^2}{\hbar e}$$

Then, the magnetic flux will be given as,

$$\Phi = BA = \frac{4\pi Mmc^2 R^2}{\hbar e}$$

$$(e)$$

Substituting Equation (e) and (b) into (a) one obtains the familiar law

$$F = \frac{e^2}{4\pi\varepsilon_0 R^2}$$

We have recovered Coulomb's law of electromagnetism practically form first principles. Therefore in principle, every external force which attempts to separate particles and antiparticles, may convert a virtual pair into a real one. If it is always an attractive force, as commonly believed today, gravity can't separate particles and antiparticles. Hence, the conjectured gravitational repulsion between matter and antimatter is a necessary condition for separation of particles and antiparticles by a gravitational field and consequently for the creation of particle-antiparticle pairs from the quantum vacuum. But while an electric field can separate only charged particles, gravitation as a universal interaction might create particle-antiparticle pairs of both charged and neutral particles. Thus, the hypothesis of antigravity opens possibility for a gravitational version of the Schwinger mechanism.

In conclusion, gravity is not an entropic force but rather a quantum force stemming from the quantum fluctuation of particles in the vacuum. That is the main statement, which is new and has not been made before. If true, this should have profound consequences.

10 SINGULARITY AVOIDANCE, THE INFORMATION PARADOX, AND PLANCK STARS

(a) Resolution of black hole singularity and the information paradox problem

The appearance of singularities in any physical theory is an indication that either something is wrong or we need to reformulate the theory itself. Singularities are like dividing something by zero. One such theory plagued by singularities is the General theory of relativity (GR) and the problems in GR arise from trying to deal with a universe that is zero in size (infinite densities). However, quantum mechanics suggests that there may be no such thing in nature as a point in space-time, implying that space-time is always smeared out, occupying some minimum region. The minimum smeared-out volume of space-time is a profound property in any quantized theory of gravity and such an outcome lies in a widespread expectation that singularities will be resolved in a quantum theory of gravity. This implies that the study of singularities acts as a testing ground for quantum gravity.

Loop quantum gravity (LQG) suggests that singularities may not exist. LQG states that due to quantum gravity effects, there must be a minimum distance beyond which the force of gravity no longer continues to increase as the distance between the masses become shorter or alternatively that interpenetrating particle waves mask gravitational effects that would be felt at a distance. It must also be true that under the assumption of a corrected dynamical equation of LQ cosmology and brane world model, for the gravitational collapse of a perfect fluid sphere in the commoving frame, the sphere does not collapse to a singularity but instead pulsates between a maximum and minimum size, avoiding the singularity.

Additionally, the information loss paradox is also a hot topic of theoretical modeling right now because it suggests that either our theory of quantum physics or our model of black holes is flawed or at least incomplete. and perhaps most importantly, it is also recognized with some prescience that resolving the information paradox will hold the key to a holistic description of quantum gravity, and therefore be a major advance towards a unified field theory of physics.

The paradox, as formulated, arises from considerations of the ultimate fate of the information that falls into a black hole: does it disappear as it falls into the black hole singularity? As well, what happens to the information of a black hole when it evaporates to nothing due to Hawking radiation? If a black hole loses all of its energy, then all of the information about all of the particles that fell in it would be lost as well. Of course the disappearance of information would be a violation of conservation laws of energy, which states that no energy or information can be destroyed.

The resolution of classical singularities under the assumption of a maximal acceleration has been studied using canonical methods for Rindler, Schwarzschild, Reissner-Nordstrom, Kerr-Newman and Friedman-Lemaitre metrics.

To reconcile quanum mechanics with general relativity, we develop a quantum geometry in relativistic phase space (Rindler space) in which the maximal (proper) acceleration of a particle is modified to read,

$$a = \frac{c^2}{2r}\alpha^n$$

Where, c is the constant speed of light, r is the linear dimension of a particle , α is the coupling constant (or size of the extra dimensions), n is a positive number (or the extra dimension number and α^n is the flux in the extra dimension

This acceleration is based on an assumption, that particles are extended objects, never to be identified with mathematical points in ordinary space. This acceleration is important because it cures strong singularities that plague general relativity. This acceleration is also a straight forward consequence of our modified uncertainty relation given as,

$$\Delta p \Delta r \geq \frac{\hbar}{2}\alpha^n \quad \Delta E \Delta t \geq \frac{\hbar}{2}\alpha^n$$

,

Where r represents the size of a star, in this case-horizon radius, p is the momentum of a particle approaching or falling into the hole of a star, α is the coupling constant and n is positive.

From the above given uncertainty principle, we derive the planck length. such that when the momentum $\Delta p = mc$, the gravitational coupling constant for gravitational interactions is $\alpha = \frac{Gm^2}{\hbar c}$ and finally n=1/2. We get the planck length as the minimum length of space-time as,

$$\Delta r = \left(\frac{\hbar G}{4c^3}\right)^{1/2}$$

(b) Planck stars

To resolve the black hole singularities and the information paradox. We consider the possibility that the energy of a collapsing star and any additional energy falling into the hole could condense into a highly compressed core with density of the order of the Planck density. If this is the case, the gravitational collapse of a star does not lead to a singularity but to one additional phase in the life of a star: a quantum gravitational phase where the gravitational attraction is balanced by a quantum pressure.

Since the energy density or pressure is expressed as force per unit surface area of a star we have,

$$\rho = \frac{F}{A}$$

Therefore nature appears to enter the quantum gravity regime when the energy density of matter reaches the Planck scale. The point is that this may happen well before relevant lengths become planckian. For instance, a collapsing spatially compact universe bounces back into an expanding one. The bounce is due to a quantum-gravitational repulsion which originates from the modified Heisenberg uncertainty, and is akin to the force that keeps an electron from falling into the nucleus. And from the uncertainity principle, this repulsion force is given by,

$$F = \frac{mc^2}{2r}\alpha^n$$

Therefore bounce does not happen when the universe is of planckian size, as was previously expected; it happens when the matter energy density reaches the Planck density in this way,

Let the surface area of a star be, $A = 4\pi r^2$ then the matter energy density will be given as,

$$\rho = \frac{mc^2}{8\pi r^3}\alpha^n$$

For a Schwarzschild black hole with radius $r = \frac{2GM}{c^2}$ and $\alpha = \frac{Gm^2}{hc}$. We have a maximum energy density value wnen n=1 given as,

$$\rho = \frac{c^7}{\pi h(8G)^2}$$

At this energy density, a Planck star is formed. The key feature of this theoretical object is that this repulsion arises from the energy density, not the Planck length, and starts taking effect far earlier than might be expected. This repulsive 'force' is strong enough to stop the collapse of the star well before a singularity is formed, and indeed, well before the Planck scale for distance. Since a Planck star is calculated to be considerably larger than the Planck scale for distance, this means there is adequate room for all the information captured inside of a black hole to be encoded in the star, thus avoiding information loss.

The analogy between quantum gravitational effects on

Cosmological and black-hole singularities has been exploited to study if and how quantum gravity could also resolve the $r = 0$ singularity at the center of a collapsed star, and there are good indications that it does. For example, from the modified uncertainty principle, when the momentum of a particle or matter falling into a black hole is Planckian $p = m_{Pl} c$ where m_{Pl} is the Planck mass, we have,

$$r = \alpha^n l_p$$

Where l_p is the Planck length. Taking $\alpha = \frac{Gm^2}{\hbar c} = \left(\frac{m}{m_{pl}}\right)^2$ we have the size of a star as,

$$r = \left(\frac{m}{m_{pl}}\right)^{2n} l_p$$

Where m is the mass of the star and n is positive. For instance, if n = 1/6, a stellar-mass black hole would collapse to a Planck star with a size of the order of 10^{-10} centimeters. This is very small compared to the original star in fact, smaller than the atomic scale but it is still more than 30 orders of magnitude larger than the Planck length. This is the scale on which we are focusing here.

The main hypothesis here is that a star so compressed would not satisfy the classical Einstein equations anymore, even if huge compared to the Planck scale. Because its energy density is already planckian.

11 BLACK HOLE ENTROPY AND HAWKING RADIATIOM

The development of general relativity followed a publication of acceleration under special relativity in 1907 by Albert Einstein. In his article, he argued that any mass will "Distort" the region of space around it so that all freely moving objects will follow the same curved paths curving toward the mass producing the distortions. Then in 1916, Schwarzschild found a solution to the Einstein field equations, laying the groundwork for the description of gravitational collapse and eventually black holes.

By definition, a black hole is an astronomical object with a very strong gravitational effect, which disturbs particles across its event horizon. It is also true from the theory of general relativity, that even light cannot escape its gravitational pull. These objects have puzzled the minds of great thinkers for many years. History puts it that, they were first predicated by John Michell and Pierre-Simon Laplace in the 18th century but David Finkelstein was the first person to publish a promising explanation of them in 1958 based on Karl Schwarz child's formulations of a solution to general relativity that characterized black holes in 1916.

In 1971, Hawking developed what is known as the second law of black hole mechanics in which the total area of the event horizons of any collection of classical black holes can never decrease, even if they collide and merge. This is similar to the second law of thermodynamics which states that, the entropy of a system can never decrease.

In 1972 Bekenstein proposed an analogy between black hole physics and thermodynamics in which he derived a relation between the entropy of black hole entropy and the area of its event horizon.

$$S = \frac{Akc^3}{4Gh}$$

In 1974, Hawking predicted an entirely astonishing phenomenon about black holes, in which he ascertained with accuracy that black holes do radiate or emit particles in a perfect black body spectrum.

$$T = \frac{\hbar c^3}{8\pi GMk}$$

Hawking beautiful result raises a number of questions. First, in Hawking's derivation the quantum properties of gravity are neglected. Are these going to affect the result? Second, we understand macroscopical entropy in statistical mechanical terms as an effect of the microscopical degrees of freedom. What are the microscopical degrees of freedom responsible for the entropy? Can we derive the entropy from first principles?

This book presents a new approach to Black hole thermodynamics that is different from that given by Loop Quantum Gravity (LQG), String theory and Bekenstein-Hawking radiation theory. The major result of the book is the derivation of the Bekenstein-Hawking area entropy law from first principles using new methods with a well defined calculation where no infinities appear. As far as this book is concerned there is no other theory from which such a calculation can proceed. Hence the methods used in here are the only one from which a detailed quantum theory of gravity precedes and where the result of the Bekenstein-Hawking area entropy law can be achieved.

Proof of the bekenstein-hawking black hole entropy law from first principles

From the Clausius definition of entropy, the entropy S of a black hole is the quantity of heat or energy Q of the hole per unit temperature of a black hole T.

$$S = \frac{Q}{T}$$

To begin with, let us find the energy Q of a black hole; Let the intensity/power per unit surface area A of a black hole be given by,

$$I = \frac{Q}{At} = \frac{F^2}{4\pi\alpha h}$$

Where, t is the time taken by the mass of a Black hole to dissipate, F is the physical force of a black hole , α is the dimensionless coupling constant and $\hbar$ is the reduced Planck constant.

The importance of the above given intensity formula is that, it gives the power of the electromagnetic waves and the Stefan Boltzmann power law for thermal radiation, when the right choice of force causing the acceleration of particles in the black body cavity is chosen. For example, for the electric force, $F = E e$, where E is the electric field, e is the charge on an electron and the fine structure/ electromagnetic coupling constant $\alpha = \frac{e^2}{4\pi\varepsilon_0 hc}$, we have the formula for the **intensity of the electromagnetic radiation** as $I = \varepsilon_0 C E^2$.

Also for the thermodynamic force, $F = \left(\frac{\pi^3 \alpha}{15}\right)^{1/2}\frac{k^2 T^2}{hc}$, on substitution into the intensity formula, we obtain $I = \frac{\pi^2 k^4}{60 \hbar^3 c^2} T^4$ which is the Stefan-Boltzmann law.

Lastly, we can also derive the power of the emitted hawking radiation in the following manner,

The power is given as,

$$P = \frac{Q}{t} = \frac{AF^2}{4\pi\alpha h}$$

The Schwarzschild black hole sphere surface area of Schwarzschild radius is given by

$$A = 4\pi\left(\frac{2GM}{c^2}\right)^2 = \frac{16\pi G^2 M^2}{c^4}$$

Also the strong force of a Black hole is given as,

$$F = \frac{c^4}{8\pi G \alpha^n}$$

Where c is the constant speed of light, G is the Universal gravitation constant, α is the size of the extra dimension, n is the extra dimension number and α^n is the flux in the extra dimension. The purpose of the force given will be seen as we go on.

On substitution we have,

$$P = \left(\frac{16\pi G^2 M^2}{c^4}\right)\left(\frac{c^4}{8\pi G \alpha^n}\right)^2 \frac{1}{4\pi \alpha \hbar}$$

Let the extra dimension number be n=1/2, we then have

$$P = \left(\frac{M^2 c^4}{16\pi^2 \hbar}\right)\frac{1}{\alpha^2}$$

Taking $\alpha = \frac{GM^2}{\hbar c}$,

Where α is the gravitational coupling constant, we have the power of Black hole radiations as,

$$P = \frac{\hbar c^6}{16\pi^2 G^2 M^2}$$

But when $\alpha = \frac{960 G M^2}{\pi \ \hbar c}$, we get the final Hawking power law as,

$$P = \frac{\hbar c^6}{15360\pi G^2 M^2}$$

Energy of a Black hole

Now let the energy of a black hole be expressed as,

$$Q = \frac{AtF^2}{4\pi\alpha h}$$

On substitution of the force $F = \frac{c^4}{8\pi G\alpha}$, (at n=1), we have the energy as,

$$Q = \frac{Atc^8}{256\pi^3 hG^2\alpha^3}$$

The change in time t for black hole evaporation is proportional to a change in the mass M of a Black hole per unit strong gravitational force of a black hole.

$$\delta t = \frac{c\delta M}{F}$$

Since F is known, then we have,

$$\delta t \sim t = \frac{Mc}{F} = \frac{8\pi GM\alpha}{c^3}$$

From which on substitution the energy will be given as,

$$Q = \frac{AMc^5}{32\pi^2 hG\alpha^2}$$

For simplicity, let $\alpha = 1$, which gives

$$Q = \frac{AMc^5}{32\pi^2 hG}$$

Secondly, we derive the temperature T of a Black hole from our force formula; Let the translational thermal kinetic energy of a particle at the surface of the event horizon of a black hole of mass M be in equilibrium with the total potential gravitational energy of a black hole given as, $kT = FR$. Where k is the Boltzmann constant and $R = \dfrac{GM}{c^2}$ is the Schwarzschild radius of the event horizon of a black hole. Since $F = \dfrac{c^4}{8\pi G \alpha}$, we then have,

$$T = \frac{Mc}{8\pi k\alpha}$$

For $\alpha = \dfrac{GM^2}{\hbar c}$, we have the temperature of a black hole given by,

$$T = \frac{\hbar c^3}{8\pi GMk}$$

Substituting Q and T in the entropy formula we have,

$$S = \frac{Q}{T}$$

$$S = \frac{kAM^2c^2}{4\pi\hbar^2}$$

If we consider a tiny black hole which is literary the Planck particle, we have

$$M = \left(\frac{\pi\hbar c}{G}\right)^{1/2}$$

This implies that the mass of a black hole is not zero but the minimum mass possible is that of a Planck particle. This mass is $\sqrt{\pi}$ times larger than the Planck mass, making a Planck particle 1.772 times more massive than the Planck unit mass.

On substitution we have the entropy of a Black hole as,

$$S = \frac{Akc^3}{4G\hbar}$$

In conclusion, we have not only proved the existence of a black hole entropy which increases with the surface area A of a black hole but we have also provided evidence for the existence of Planck particles, whose mass is $\sqrt{\pi}$ times larger than the Planck mass.

OTHER APPROACHES

The First Approach

In this method we reduce the famous Einstein field equation ($G_{\mu\nu} + \Lambda g_{\mu\nu} = \frac{8\pi G}{c^4} T_{\mu\nu}$)where,the expression on the left represents the curvature of space time while the expression on the right represents the matter-energy content of the universe) to,

$$\frac{1}{R^2} = \frac{8\pi G}{c^4} P_{eg}$$

(1)

Where, R is the radius of a body of mass M, $P_{eg} = \sigma_m \frac{f_e f_g}{\hbar c}$ is the Pressure-Energy density relationship with the coupling of mass (the ratio of the atomic mass, m to the Planck mass M_{pl}) and the electric f_e and gravitational force f_g .The ratio, $\sigma_m = \frac{m}{M_{pl}}$ is introduced to correct for particles approaching the Planck length scale $m \to M_{pl}$. The total electric potential energy of a black hole from (1), is given by,

$$E_e = f_e r = \frac{\hbar c^5}{8\pi G E_g \sigma_m}$$

If we let the potential gravitational energy be, $E_g = mc^2$ then the electric energy is,

$$E_e = \frac{\hbar c^3}{8\pi G M \sigma_m}$$

Since the thermal energy is given by $E_{thermal} = kT$, where k is the Boltzmann constant, therefore by the principal of Equipartition

$$E_{thermal} \sim E_e \Rightarrow T = \frac{\hbar c^3}{8\pi GMk\,\sigma_m}$$

(2)

And when $\sigma_m = 1$, we get the usual Hawking temperature as,

$$T = \frac{\hbar c^3}{8\pi GMk}$$

(3)

We know that, entropy is energy divided by temperature.

What is the total energy of a Black hole? Assuming a law which states that the intensity of the emitted radiation increases as the square of the electric force $I = \beta F_e^2$, where the constant $\beta = \frac{1}{4\hbar}$. But we can also write the intensity in terms of energy as, $I = \frac{E_T}{tA}$, where t is time and A is the surface area of Schwarzschild black hole. The total energy of a Black hole will then be given as,

$$E_T = \frac{F_e^2 tA}{4\hbar}$$

Let the time taken by a Black hole to evaporate be $t = \frac{Mc}{F_e}$ F_e is known from (1). Since from the Newtonian law of gravitation $F_e R^2 = GM^2$ we then have the total energy of a black hole as,

$$E_T = \frac{Ac^6}{32\pi\sigma_m G^2 M}$$

(4)

Then the entropy of a Black hole is given by

$$S = \frac{E_T}{T}$$

Substituting in (4) and (3) we obtain the Bekenstein-Hawking area entropy law,

$$S = \frac{Ac^3 k}{4\hbar G}$$

The Second Approach

It is here by hypothesized that the gravitational field will create particles and emit them only if the electromagnetic force of such particles were equal to the

$$F = \frac{Me}{r}\sqrt{\frac{Gp}{2\hbar\varepsilon_0\lambda}}$$

force (unknown in literature) .Where p, is the momentum of a particle.

Then under general conditions, the force given will reduce to the Reissner-Nordstrom metric as given here, if the momentum of an electron at a distance r from the singularity point to the event horizon is related to the de Brogile

wavelength as $p = \frac{2\pi\hbar}{\lambda}$, and both the distance r and wavelength λ was the

product of the speed of light c and the period T as r=cT and $\lambda = cT$, then the force will be given by,

$$F = \frac{Mp}{r\hbar}\sqrt{\frac{Ge^2}{4\pi\varepsilon_0}}$$

But since $\frac{p}{2\pi\hbar} = \frac{1}{\lambda}$, then we have,

$$F = \frac{2\pi M}{T^2}\sqrt{\frac{Ge^2}{4\pi\varepsilon_0 c^4}}$$

This reduces to,

$$F = \frac{2\pi M}{T^2}r_q$$

where $r_q = \sqrt{\dfrac{Ge^2}{4\pi\varepsilon_0 c^4}}$ is the Reissner-Nordstrom radius of a charged black hole.

Having derived the Reissner-Nordstrom metric from our force formula, we now return to our exercise of deriving the temperature of a black hole. We consider a particle with charge e, exhibiting deBrogile wave properties of momentum and wavelength from the centre of mass M of a black hole. We then assume that this particle experiences an electromagnetic force due to the magnetic and electric field created by other particles in its surrounding area. The same particle also experiences a force due to the strong gravitational field emanating from the black hole.

Equating the two forces as,

$$\frac{Me}{r}\sqrt{\frac{Gp}{2\hbar\varepsilon_0\lambda}} = \frac{e^2}{4\pi\varepsilon_0 r^2}$$

From this expression we obtain the momentum of a particle as,

$$p = \frac{\hbar e^2 \lambda}{2\pi A \,\varepsilon_0 G M^2}$$

This is the momentum possessed by a particle (emitted by the gravitational field of a black hole) at the surface of the event horizon, where $A = 4\pi r^2$ is the spherical surface area of the horizon.

For relativistic effects, the kinetic energy of a particle will be related to its momentum by **K.E=pc** and to the Boltzmann's law by **K.E=kT**, where **k** is the Boltzmann's constant and T is the absolute temperature. By similarity we can equate the two energies as **pc=kT**, then from the equation of momentum we can obtain the temperature as,

$$T = \frac{\hbar e^2 \lambda c}{2\pi A \varepsilon_0 G M^2 k}$$

Expressing the permittivity of free space in terms of the permeability of free space $\varepsilon_0 = \frac{1}{\mu_0 c^2}$, we obtain the Hawking temperature of a black hole as,

$$T = \left(\frac{4e^2 \mu_0 \lambda}{AM}\right) \frac{\hbar c^3}{8\pi G M k}$$

In a more general form, in terms of energies it can be expressed as,

$$T = \left(\frac{4e^2 \lambda}{A\varepsilon_0 Mc^2}\right) \frac{\hbar c^3}{8\pi G M k}$$

In an attempt to prevent the violation of the generalized second law of thermodynamics, Bekenstein proposed a universal upper bound on the ratio entropy to energy for bounded systems (Phys RevD23, 287-1981), which was later rejected by Unruh and Wald in 1982. They proposed a thought experiment in which a box lowered down into a black hole felt an effective buoyancy force which was caused by the acceleration radiation felt by the box near the black hole. They argued further that, this buoyancy force would guarantee a lower bound on the energy gain of the black hole, hence saving the generalized second law without a need for entropy bound.

In this approach we give a formula for the buoyancy force which is different from the Unruh and Wald formula which appeared in their 1982 paper. At a distance r from the center of mass m of a black hole, the buoyancy force is given by,

$$F_B = \frac{rc^6}{8G^2 m}$$

From the above force formula the energy gain by the black hole will be given by,

$$W_B = \frac{Ac^6}{32\pi G^2 m}$$

Where, A is the area of the event horizon.

Since entropy is the ratio of energy to temperature, $S_B = W_B / T_B$ and temperature of a black hole is known from equation 8, then the entropy of a black hole is given by,

$$S_B = \frac{Akc^3}{4\,Gh}\left(\frac{As_0 M c^2}{4e^2\lambda}\right)$$

The expression in brackets, i.e $\frac{4e^2\lambda}{As_0 M c^2}$ appearing in the above formula, is a dimensionless number.

The Third Approach

A Black hole is a mathematically defined region of space time exhibiting such a strong gravitational pull that no particle or electromagnetic radiation can escape from it. Many theories have been created to explain the properties of the black hole but the theory created here is far more different from the other theories although it may give the same results. Using a quite different approach towards solving a problem is efficient since it comes with it new predictions in the process which could have been hidden in other approaches. Below we try to present adhoc proofs-laws that may be of help in building our theory about black holes.

It is well known that the electric field is force per unit charge but here a generalized equation for an electric field created by an electron exhibiting wave properties in the nucleus of an atom in the gravitational field on a quantum scale is given by

$$E = \frac{1}{r}\sqrt{\frac{Gm^3 f}{2\hbar\varepsilon_0}}$$

$$(g)$$

Then the electric force in this case will be formulated as

$$F_1 = \frac{e}{r}\sqrt{\frac{Gm^3 f}{2\hbar\varepsilon_0}}$$

$$(h)$$

The surface area at a radius r of orbit of an electron of mass m around the nucleus of an atom in a wave like manner is given by

$$\text{surface area}(A) = \frac{\lambda\mu_0 e^2}{m}$$

$$(i)$$

The time taken by the magnetic field B of an electron to pass through a given surface is

$$time\,(t) = \frac{\lambda \varepsilon_0 AB}{e}$$

$$(j)$$

Note: the above expression is the same as Faraday's induction law.

The gravitational force acting on all matter in the universe or the modified gravitational force is given as

$$F_2 = \left(\frac{Gm^3}{r^2}\right)\left(\frac{e}{2B\lambda\hbar\varepsilon_0}\right)$$

$$(k)$$

The above formulas are important in deriving the formula for the temperature, entropy and the time taken by a black hole to evaporate as shown below;

It is known that the kinetic energy KE of molecules in the Boltzmann hypothesis is related to the temperature of the body in question in this case a black hole (in relation to the black body) by $KE = \varphi T$ where φ is Boltzmann's constant. The formula for the kinetic energy can be derived by using a hypothesis that the electromagnetic force – coulombs force is equal to eqn(h) as

$$\frac{ke^2}{r^2} = \frac{e}{r}\sqrt{\frac{Gm^3 f}{2\hbar\varepsilon_0}}$$

On squaring both sides of the equation, cancelling like terms and taking into account that the frequency of an electron is $f = \frac{v}{\lambda}$, then the kinetic energy of an electron inside the black hole is given by

$$KE = \frac{\lambda \mu_0 e^2}{A} \frac{c^3 \hbar}{8\pi G m^2}$$

Since the surface area is given as from eqaution(i) then the kinetic energy of molecules or particles (for an ideal gas) within the black hole will be given by

$$KE = \frac{c^3 \hbar}{8\pi G m} = T\varphi$$

$$(1)$$

Then from Boltzmann's relationship the temperature of the black hole is formulated as

$$T = \frac{c^3 \hbar}{8\pi G m \varphi}$$

$$(m)$$

By definition entropy is a measure of disorder. To solve the entropy of black holes we shall consider a very complex argument about the entropy in question. We assume that the modified gravitational force given by equation (k) is identical to the modified electric field given by equation(h) as,

$$\left(\frac{Gm^3}{r^3}\right)\left(\frac{\circ}{2B\lambda \hbar s_0}\right) \equiv \frac{\circ}{r}\sqrt{\frac{Gm^3 f}{2\hbar s_0}}$$

Then squaring both sides of the equation and multiplying through by Gc^5 one obtains a new relation of forces on both sides given as

$$\frac{tc^7}{16\pi G^2 m} = \frac{Ac^6}{32\pi rm\, G^2}$$

Both the left and right hand side represent a force. From the left hand side t is the expression of time given by,

$$t = \frac{\hbar e^2}{2m^3 c^2 G \varepsilon_0}$$

Since the heat is the product of the force on a particle and the distance r from the centre of the black hole, then using the force on the right hand side of the above equation the heat will be given by

$$Q = \frac{Ac^6}{32\pi m\, G^2}$$

Remember the temperature of the black hole is also known from equation (m) and by definition the entropy of the system is the change in heat per unit temperature $\frac{Q}{T}$, then the entropy of the black hole will be given by

$$S = \frac{A\varphi c^3}{4G\hbar}$$

This implies that the entropy of a black hole is proportional to its surface area.

The time taken by a black hole to evaporate

Assuming that particles that formed a black hole are moving away or are separating from it after a given time of its existence, if we measure the relative speed of these particles in relation to the energy they carry we obtain a relationship given by

$$\frac{v^2}{c^2} = \frac{8\pi G}{c^2}\left(\frac{W}{8\pi r}\right)$$

(o)

Where v is the velocity of these particles as measured relative to the speed of light c and W is the energy carried by the particles as they move away from the centre of the black hole at a distance r.

If we let the force causing the particles to separate from the black hole be given as, $\frac{Gm^3_e \; v}{2r\lambda B\hbar\varepsilon_0 \; c}$, then the energy of these particles will be given by

$$W = \frac{Gm^3_e \; v}{2r\lambda B\hbar\varepsilon_0 \; c}$$

Substituting this in equation (o), we obtain a relationship of time as given by the law 3 of equation (j) as

$$t = \frac{v^2}{c^2}\left(\frac{\pi G^2 m^3}{\hbar c^4}\right)$$

The velocity of the particles in the astronomical lab will be measured as v= 4.193E6 m/s and since the speed of light is a constant then the time taken by a black hole to evaporate is given by,

$$t = \frac{5120\pi G^2 m^3}{\hbar c^4}$$

12 THE PRINCIPLE OF LEAST ACTION

In this section a new approach towards Quantum Gravity is presented, we try to study the theory from new assumptions which are far more different from models that have been used by scientists for centuries. Most physicists have clung to old models or complex mathematical scientific methods to explain phenomenon. They are trying to explain physics using the mathematics that was earlier used by Einstein, Richard P. Feyman etc. The mathematical ideas that were presented by these physicists were complicated and such has been difficult to understand and of course misleading (Read Lost in Math, a book by Sabine Hossenfielder). For example, in a statement by Dr Lee Smolin, "the mathematisation of physics has resulted in the reduction of the cosmos to a mathematical entity, which has not only confused physicists but accounts for their worst and most distracting assertions".

There is a wide spread speculation that the mathematical formulation of physics has not only confused physicists but has also lead to failures in the development of a quantum theory of gravity.

Physics as a subject should be simple and elegant, trying to explain everything from one source. In other words trying to explain all of physics from one equation call it "**the principle of least action**". Imagine deducing the equations of gravity, quantum mechanics, electromagnetism, heat etc from one equation, wouldn't it be unique than holding about ten books about a different subject of physics each starting from its own source?

The principle of least action in simple terms means; to understand how to get from point A to point B using the least amount of physical work for example taking an elevator rather than using the stairs, in otherwords deducing the most fundamental physical equations from one principle as we are yet to find out.

Assuming that the ratio of the gravitational force to the electric force is equal to the gravitational coupling constant, we then have,

$$[8\pi G/c^4][Ee][GM^2/c]=n^2\hbar \ 1$$

Where G– gravitational constant, n- quantum number, c- constant speed of light, e- charge on an electron, M – mass and is the reduced Planck constant.

Ee is the electromagnetic force, and $8\pi G/c^4$ is the gravitational force at the swcharzichild's radius. From the above given principle, we deduce the temperature of a black hole, the time taken by a black hole to evaporate, entropy of a black hole, the wiedemann franz law and the stefan's radiation law.

(a) The temperature of a black hole

On arranging equation one to get the random translational kinetic energy, we obtain

$$[GM/c^2]Ee = n^2 c^3 \hbar /8\pi\, GM = kT$$

Where k is the boltzmann's constant and T is the temperature. Hence at n=1,

$$T = c^3 \hbar /8\pi\, GMk\, 2$$

This is the known temperature of a black hole that was originally derived by Hawking

(b) Time taken by a black hole to evaporate

On dividing through eqn1 by the momentum Mc we obtain, the time t given by

$$t = Mc/Ee = 8\pi G^2 M^3 /n^2 \hbar c^4$$

Such that when n=0.03953

$$t = 5120\pi G^2 mo^3 /\hbar c^4\, 3$$

This is the known time of a black hole that was originally derived by Hawking

(c) Entropy of a black hole

Squaring both sides of equation 1 and arranging we generate the intensity as

$$W/tA = E^2 e^2 /2n\hbar = n^3 c^{10} \hbar /256\pi^3 G^4 M^4\, 4$$

Where A is the area on which the radiations fall, W is energy, and t is time. But entropy is energy divided by temperature Eq so then

$$W/T = S = (n^3 c^{10} \hbar /256\pi^3 G^4 M^4)(tA/T)$$

Since t is known from Eq3 and T from Eq2, then at n= π

$$S = Akc^3 /4G\hbar$$

This is the known Bekenstein-Hawking area entropy law

(d) Thermal properties of solids

From the intensity equation4,

$$E^2 e^2 /2n\hbar = c^{10} \hbar /256\pi^3 G^4 M^4$$

Arranging the above equation to introduce in the translational kinetic energy obtained above [kT= $c^3 \hbar /8\pi$ GM], we have

$$\pi M^2 G^2 E^2 /3c^4 = (\pi^2 /3e^2)(c^3 \hbar /8\pi GM)^2 = (\pi^2 /3e^2)T^2 k^2$$

Dividing both sides by T we obtain on the left hand side of the equation the ratio of thermal conductivity K to electric conductivity δ as

$$K/\delta = (\pi^2 /3)(k/e)^2 T$$

This is the known **wiedemann fanz law**

(e) Stefan's Radiation Law

Still from the intensity Eqn4 we can arrange the expression on the left hand side of the equation, to read as

$$W/tA = E^2 e^2 /2n\hbar = (1.875n^3 /\pi^2)(\pi^2 /60\hbar^3 c^2)(c^3 \hbar /8\pi GM)^4$$

This is the same as Eqn4 only that it is arranged to predict something. But at n=1 and $(c^3 \hbar /8\pi$ GM$)$ =kT, the rate at which energy is radiated is given by

$$W/t = A(1.875/\pi^2)(\pi^2/60h^3c^2)(Tk)^4 = 0.19\sigma AT^4$$

Where $\sigma = \pi^2/60h^3c^2$ is the Stefan boltzmann's constant

In conclusion, I encourage further research into this field. In other words this could be a stepping stone towards the development of a theory of everything via a simpler path of least action.

PARTIII BEYOND THE STANDARD MODEL

13 THE COSMOLOGICAL CONSTANT PROBLEM

It is known that in a homogenous cosmological universe, a positive cosmological constant induces repulsive forces. The question; is there a classical formula of the force of the cosmological constant like that of the gravitational force? How does the repulsive force relate to the cosmological constant and the coupling constant? How does understanding the energy density in relation to force, change the way we perceive Einstein's field equation? The section sets out to answer these and more questions about the cosmological constant problem. (see chapter 26 on the origin of gravity and electrodynamics)

Dr Lee Smolin represents the perimeter institute for theoretical physics. He claims that the mathematisation of physics has resulted in the reduction of the cosmos to a mathematical entity, which has not only confused physicists but accounts for their worst and most distracting assertions.

There is a wide spread speculation that the mathematical formulation of physics has not only confused physicists but has also lead to failures in the development of a quantum theory of gravity.

Although both general relativity and quantum mechanics work well in the domain of their applicability, it's unfortunate that there is no unified theory of gravity with quantum mechanics.

It is proposed that the unification of gravity with quantum mechanics will require us to change the kind of mathematics that was used by either Einstein or Schrödinger etal.. in the development of both theories. But why do we bother at all if there is another way in which we can express the theory better without the use of tensor fields.

The problem with the mathematical formulation of general relativity if at all it exists stems from the non existence of its experimental observation which wasn't the case with quantum mechanics. The formulation of quantum mechanics was based on the existence of experimental observations. Therefore quantum mechanics was founded on the existence of experiments which wasn't the case with general relativity. Einstein had to base his theorization on thought experiments which could or wouldn't be nearer to any experimental confirmation of the phenomenon being studied.

The same is also true for the formulation of quantum gravity. There is no sound experimental proof for the existence of quantum gravitational effects and therefore scientists like Hawking Stephen have also clung to the old formulations that were used by Einstein and his contemporaries to develop a quantum theory of gravity.

In this brief notice we show that an existence of a unified theory is rooted deep into the unnoticed pressure-energy density similar to the stress energy tensor appearing in Einstein field theory. Our major aim therefore is to provide proof for the questions set out below;

(i) If the cosmological constant introduces a force of repulsion between bodies. Is it true that the force increases in simple proportion to the cosmological constant and the coupling constant?

(ii) Is there a classical formula of the force of the cosmological constant like that of the gravitational force?

Einstein's general relativity equations famously described the curvature of space-time as the mechanism for gravity. In the original theory, Einstein added a "cosmological constant" that acted as an expulsive force to counteract gravity. That stabilized the universe so it didn't collapse in on itself, but Einstein abandoned the idea when further astronomical observations showed the universe was accelerating and not static, as the great physicist had thought.

Analogous to the known Einstein field equation, the curvature of space (cosmological constant) is here related to the energy density $^{\omega}$ as,

$$\Lambda = \kappa\omega = \kappa\left(\frac{F^2}{8\pi\alpha hc}\right)$$

$$(1)$$

Where $\kappa = \dfrac{8\pi G}{c^4}$ a constant appearing in Einstein's field equation, F is is the force in an interaction and α is the coupling constant.

The above expression implies that the cosmological constant is related to the force and therefore increases as a square of the force.

For the energy density in electric field, where $F = Ee$ and $\alpha = \dfrac{e^2}{4\pi\varepsilon\hbar c}$, the energy density will be given by, $\omega = \dfrac{F^2}{8\pi n\hbar c} = \dfrac{\varepsilon E^2}{2}$.

While for the energy density in the gravitational field, where $F = mg$ and $\alpha = \dfrac{Gm^2}{\hbar c}$, the energy density will be given by, $\omega = \dfrac{F^2}{8\pi n\hbar c} = \dfrac{g^2}{8\pi G}$. This can be written in simple terms as $\omega = \dfrac{\eta g^2}{2}$, where $\eta = \dfrac{1}{4\pi G}$.

From (1) therefore, the force responsible for the expansion of the universe is related to the cosmological constant by,

$$F = E_{pl}(\alpha\Lambda)^{1/2}$$

$$(2)$$

Where $E_{pl} = 1.9605 \times 10^9 J$ is the Planck energy.

Given the Planck (2015) values of $\Omega_\Lambda = 0.6911 \pm 0.0062$ and $H_0 = 67.74 \pm 0.46$ (km/s)/Mpc $= (2.195 \pm 0.015) \times 10^{-18}$ s^{-1}, Λ has the value of $1.11 \times 10^{-52} m^{-2}$ as given in wikimedia commons.

Based on the above given value, the force will then have a value of

$$F_{ob} = 2.0655 \times 10^{-17}(\alpha)^{1/2}$$

$$F_{ob} \sim 1.8 \times 10^{-18} N$$

This therefore is a force responsible for the expansion of the Universe. It is such a small force that will require sophiscated machines to measure. While the above force value is based on the fine structure constant, there is a value that is even smaller than that value by, $F_{ob} \sim 1.58 \times 10^{-36} N$ at $\alpha = 5.87 \times 10^{-39}$ between two protons.

However in quantum electrodynamics (QED) we compute a much larger value of $F_{QED} \sim 2.82 \times 10^{44} (\alpha)^{1/2} N$. This huge discrepancy is known as the cosmological constant problem. Therefore the relative strength of the force will be given by;

$$\frac{F_{QED}}{F_{ob}} \sim 10^{61}$$

The above value is in agreement with the Hubble age to the Planck time, which is the same as the total mass of the universe to the Planck mass as,

$$\frac{F_{QED}}{F_{ob}} = \frac{t_H}{t_{pl}} = \frac{M_U}{M_{pl}} \sim 10^{61}$$

The above given relationship implies a persistence constant error that is evident when comparing observational and theoretical calculations. This error needs to be distributed uniformly in order to correct for large discrepancies which accrue to calculated values in relation to observed values.

The problem lies in knowing the observed force value to the calculated value, since the force ratio doesn't correspond to the other ratios of time and mass. In other words changing the ratio $\frac{F_{QED}}{F_{ob}}$ to $\frac{F_{ob}}{F_{QED}}$ will cause other ratios to change.

It is therefore observed that the ratio of Hubble age to the Planck time and the total mass of the universe to the Planck mass will only be in line or tally with the Planck force to the Hubble force by a value $\sim 10^{61}$ and not otherwise.

Keeping other factors constant it is clear from the above given observations that the mysterious, repulsive force pulling galaxies apart is proportional to the coupling constant value in a given interaction. This proposal will be of such a great importance to the work of researchers involved in the field of quantum gravity

The Big Problem: *Why does the Zero-Point Energy Of the Vacuum not cause a Large Cosmological Constant? What Cancels it out?*

The cosmological constant problem is the large discrepancy between the experimental observed value of the cosmological constant and its theoretical calculated value. Whereas observations give values of the energy density and the cosmological constant as,

$$\omega \approx 1.18 \times 10^{-9} \, J/m^3 \quad \Lambda \approx 1.11 \times 10^{-52} m^{-2}$$
$$, \qquad\qquad\qquad (a)$$

The calculated theoretical values of the energy density and the cosmological constant are larger by,

$$\omega \approx 2.531 \times 10^{114} \, J/m^3 \quad \Lambda \approx 5.23 \times 10^{71} m^{-2}$$
$$, \qquad\qquad\qquad (b)$$

The question is; why does the Zero-point energy of the vacuum not cause a large Cosmological constant? What cancels it out? Before we proceed, let me show you how the above given values for the energy density and cosmological constant come about.

Consider an expanding universe, where a test particle of mass m_i in this universe subject to a force F is accelerating with a constant acceleration a. The energy density ω in this case is related to the force by,

$$\omega = \frac{F^2}{8\pi\alpha hc}$$

Where, $F = m_i a$, α is the coupling constant. This then gives,

$$\omega = \frac{m_i^2 a^2}{8\pi\alpha hc}$$

(1)

Also, the cosmological constant is related to the acceleration by,

$$\Lambda = \frac{a^2}{\alpha c^4}$$

(2)

To obtain the experimental observed values of energy density and the cosmological constant, we assume the following values of mass and acceleration

$m_i = 2.18 \times 10^{-8} kg$ - Planck mass

$a = 1.2 \times 10^{-10} m/s^2$, small acceleration (MOND by Milgrom)

$\alpha = 1/137$, Fine structure constant

Substituting these values in (1) and (2) we surely get values in (a)

To obtain the theoretical values of energy density and the cosmological constant, we assume the following values of mass and acceleration

$m_i = 2.18 \times 10^{-8} kg$, Planck mass

$a = 5.5608 \times 10^{51} m/s^2$, maximal acceleration (Planck units)

$\alpha = 1/137$, Fine structure constant

Substituting these values in (1) and (2) we surely get values in (b)

We observe that, the common factor in the above calculations is the Planck mass and the coupling constant. This implies that, to solve the cosmological constant problem we need to create a relationship between the energy density and the cosmological constant where the acceleration and coupling constant disappears as,

$$\Lambda = \frac{8\pi h}{m_i^2 c^3}\omega$$

$$(3)$$

$$\Lambda = \frac{9.82 \times 10^{-59}}{m_i^2}\omega$$

This equation means that, the zero point energy of the vacuum will not cause a large cosmological constant because of the inverse square law given above. This means that, the large mass of the interacting particle cancels out the large cosmological constant caused by the zero point energy density of the vacuum.

For a small interacting particle $m_i \approx 10^{-70} kg$, we get a large cosmological constant and vice versa. This means that, all observations for the cosmological constant assume the interacting particle to be the Planck mass. Implying that the Planck mass is the least mass responsible for the acceleration of the universe, below the Planck mass, the universe becomes static. Therefore the above equation can be used to determine the mass of a particle responsible for the expansion of the universe by assuming a constant cosmological constant.

(a) Proof of Newton's Law of Universal Gravitation

Previously we showed that the energy density ω is related to the force F by,

$$\omega = \frac{F^2}{8\pi\alpha hc}$$

Where,ℏ is the reduced Planck constant, c is the constant speed of light and is the coupling constant or principal quantum number

In this short notice we clearly prove that F is the gravitational force that was put forward by Newton as we are yet to see below,

Independent of the mass and distance, the force between two particles as in the case of the Casmir affect is therefore given by,

$$F = \sqrt{8\pi\alpha\hbar c\omega}$$

(1)

Below I give two conditions on which the above force will reduce to the Newton's universal law of gravitation

The energy density is related to the cosmological constant by,

$$\omega = \frac{m^2 c^3}{8\pi\hbar}\Lambda$$

This was previously derived and m was the Planck mass

Here m is taken to represent the mass of the particle in circular orbit around a massive body of mass M at a distance or radius of curvature R from M. Where,

$$\Lambda = \frac{1}{R^2}.$$

The coupling constant or principal quantum number is here given as a ratio of the areas as,

$$\alpha = \frac{A_s}{A}$$

Where $A_s = \frac{4\pi G^2 M^2}{c^4}$, is the Schwarzschild area occupied by a massive body and $A = 4\pi R^2$ is the total area of circular orbit of a mass m.

Substituting condition 1 and 2 into equation (1) above we obtain the Newton's law of gravitation as,

$$F = \frac{GMm}{R^2}$$

This derivation is proof that gravity is a result of the quantum vacuum energy density. While this has proved to be a short insight into the emergency of gravity, we shall have a long discussion of this research in the coming chapters.

14 MASS AIN'T WHAT IT USED TO BE

The origin of mass problem is at the forefront of those big unsolved problems in the standard model of physics. My first insight about mass came in 2000 in a lecture about Newton's mechanics probably about the study of Newton's second law of motion. The problem is important to me because the primary role of mass is to mediate gravitational interaction between bodies, and no theory of gravitational interaction reconciles with the currently popular standard model of particle physics. But because the problem started with inertia, we again revisit Newton's law to create a model through which all the masses of elementary particles can be generated.

Recall from the previous chapter the modified Rindler space with an acceleration given by,

$$a = \frac{c^2}{r}\alpha^{1/2}$$

Because we desire to have a mass formula just in terms of universal interaction couplings (like e) and basic constants like G and $\hbar$. We are therefore propted to use the radius of a RN electrically charged black hole as this is purely made of only universal constants.

$$r = \left(\frac{Ge^2}{4\pi\varepsilon_0 c^4}\right)^{1/2}$$

Then inserting this radius into the formula for acceleration we obtain,

$$a = \frac{c^4}{e}\left(\frac{4\pi\varepsilon_0\alpha}{G}\right)^{1/2}$$

Where, c is the constant speed of light, e is the charge on an electron, α is the dimensionless coupling constant, ε is the permittivity of free space and G is the universal gravitational constant.

We have therefore obtained the acceleration formula made of purely universal fundamental physical constants. Then from Newton's law of motion where force is the product of mass and acceleration, F=ma, we have

$$F_N = \frac{mc^4}{e}\left(\frac{4\pi\varepsilon_0\alpha}{G}\right)^{1/2}$$

Having found the force on a particle due to its inertia, we would like also to deduce the force of attraction on the same particle. Assuming the self attraction of a particle of radius R to be caused by the quantum fluctuation of the vacuum, then the Casimir force will be given by,

$$F = \frac{\hbar c}{R^2}$$

Taking the radius of the particle to be of Schwarzischild radius $R = \frac{Gn}{c^2}$, then the force due to self attraction will be given by,

$$F_c = \frac{\hbar c^5}{G^2 m^2}$$

By the principle of equivalence $F_N = F_c$, this then gives the mass formula as,

$$m = \left(\frac{\hbar^2 c^2 e^2}{4\pi\varepsilon_0 G^3 \alpha}\right)^{1/6}$$

We have therefore created a mass formula that is made of purely fundamental physical constants. The next step is to test the theory to see if it actually makes fundamental sound predictions. First we would like to know if the above given mass formula gives the Planck mass value and what coupling constant makes this possible?

We find that, when the dimensionless physical constant is the fine structure constant or the electromagnetic coupling constant $\alpha_e = \frac{e^2}{4\pi\varepsilon_o\hbar c}$, then the mass is probably the planck mass,

$$M_{pl} = \left(\frac{\hbar c}{G}\right)^{1/2}$$

Because the above result is general and true on cosmological grounds, then the other masses of elementary particles will be generated in a similar manner but on grounds that the Planck mass is the upper bound on mass. For example, for the case of the gravitational coupling constant, $\alpha = \frac{Gm^2}{\hbar c}$

$$m = \left(\frac{(\hbar c)^3 e^2}{4\pi\varepsilon_o G^4}\right)^{1/8} = 0.54M_{pl}$$

However things become complicated with the Higgs mass. The coupling constant required in calculating the Higgs mass of $2.2375 \times 10^{-25} kg$ according to the theory given above is enormous with a value given as $\alpha = 6.2 \times 10^{99}$. To show that the result given here is correct, we embark on the derivation of the life time of a star as,

From the modified uncertainity principle given in chapter1, the life time of the Higgs mass is here given by,

$$\Delta t = \frac{\hbar}{2\Delta E}\alpha^{1/2}$$

Where ΔE is the change in energy of the particle and according to Einstein mass-energy relation this energy is given as,

$$\Delta E = mc^2 = m = \left(\frac{h^2 c^{14} e^2}{4\pi\varepsilon_0 G^3 \alpha}\right)^{1/6}$$

For the Higgs boson particle, where the coupling $\alpha = 6.2 \times 10^{99}$, the binding energy will be calculated to be,

$$\Delta E = 2.0136 \times 10^{-8} J$$

Then on substitution into the life time formula above, we get the life time of the Higgs boson as,

$$\Delta t = 2.063 \times 10^{23} s$$

The above calculated lifetime agrees with experimental observations. This therefore proves that the value of the coupling constant for the higgs mechanism given here is correct.

Last but not least, when the dimensionless coupling constant is unity, that is $\alpha=1$. We obtain the following mass value,

$$m = 0.44 M_{pl}$$

In general, the coupling constant and generation of mass must follow this simple rule

$$\alpha = \left(\frac{M_{pl}}{m}\right)^6 \alpha_e$$

Such that when m=1.4Mpl, we get the value of the strong interaction gluon coupling (asymptotic freedom) as,

$$\frac{\alpha}{\alpha_e} = 0.13$$

We have shown above that the generation of mass is possible however it requires the determination of the exact number of the dimensionless coupling constant of the underlying theory. We have found that the coupling constant in the generation of the higgs mass is enormous and requires a small energy to probe it. We have also shown that the particles collapse in the same way as the stars do which calls for another branch of physics to probe this study efficiently. Therefore the study of the origin of mass is important in the study of the origin of the universe and the unification of all physics using the coupling constants for different fundamental laws.

15 WHAT IS THE RADIUS OF A PROTON?

Today the proton radius is measured via three methods that is, the spectroscopy, nuclear scattering and muonic hydrogen (2010 experiment) methods.

The spectroscopy method uses the energy levels of electrons orbiting the nucleus. This method produces a proton radius of about 8.768×10^{-16} m, with approximately 1% relative uncertainty.

The nuclear method is similar to Rutherford's scattering experiments that established the existence of the nucleus. Small particles such as electrons can be fired at a proton, and by measuring how the electrons are scattered, the size of the proton can be inferred. Consistent with the spectroscopy method, this produces a proton radius of about 8.775×10^{-16} m.

The muonic hydrogen 2010 method by Pohl et al. is similar to the spectroscopy method. However, the much higher mass of a muon causes it to orbit 207 times closer than an electron to the hydrogen nucleus, where it is consequently much more sensitive to the size of the proton. The resulting radius was recorded as 8.42×10^{-16} m. This newly measured radius is 4% smaller than the prior measurements, which were believed to be accurate within 1%.

The discrepancy between the measured values of the proton radius by the methods given above is what is called the proton radius puzzle and the discrepancy might be due to new physics, or the explanation may be an ordinary physics effect that has been missed.In what follows, I deduce the spectroscopy and muonic radius of the proton and thereafter provide a reason for the discrepancy.

In Newtonian law of motion for a body to orbit around another body, there must be a centripetal acceleration $\frac{v^2}{R}$ to keep the body in orbit. Where v is the velocity and R is the distance from the center. In the Hydrogen atom this is different the acceleration is given by,

$$a = \frac{c^2}{R_t}$$

Where c is the constant speed of light and $R_t = \frac{l_p^2}{d}$

l_p^2 is the planck area and d is the Schwarzschild radius of the proton (the radius of the event horizon of a proton black hole).

Then the acceleration of an electron will follow the inverse square law given by,

$$a = \frac{dc^2}{l_p^2}$$

We know that the total electric potential energy of an atom is given by,

$$V = \frac{e^2}{4\pi\varepsilon_o d}$$

Eliminating d from the potential we have,

$$V = \frac{e^2 c^2}{4\pi\varepsilon_o a l_p^2}$$

We have therefore created the potential that falls off as the acceleration.

By measuring the energy released when the excited electrons fell back to lower-energy states, the Rydberg constant could be calculated, and from this the proton radius inferred.

Since the energy of the photon $E = \frac{2\pi hc}{\lambda}$ released is equal to the total potential energy of an atom we have the radius of the proton as,

$$r_p = \frac{\lambda}{2\pi} = \frac{l_p^2 a}{c^2 \alpha_e}$$

Where, λ is the wavelength and $\alpha_e = \frac{e^2}{4\pi\varepsilon_0 hc}$ is the electromagnetic coupling constant.

Assuming an acceleration that was earlier given in chapter 14 of,

$$a = \frac{2nc^2}{R_s}$$

Where n is positive coupling number and $R_s = \frac{2Gm_p}{c^2}$ is the proton Schwarzschild radius. Then on subsitution into the proton radius formula we have,

$$r_p = \frac{2nl_p^2}{\alpha_e R_s}$$

From the above formula we deduce that,

(i) When n=0.03038 for electrons orbiting the nucleus, we get a proton radius of about 8.768×10^{-16} m, which is the exact proton radius that was produced by the spectroscopy method.

(ii) When n=0.0292 for electrons orbiting the nucleus, we get a proton radius of about 8.42×10^{-16} m, which is the exact proton radius that was produced by the 2010 experiment by Pohl et al.

We notice that the difference between the two values of n for the methods given above is the anomalous magnetic dipole moment given by,

$$\Delta n = n_1 - n_2$$

$\Delta n = 0.03038 - 0.0292$

$\Delta n = 0.00118$

Therefore the discrepancy in the radius of the proton by the two methods is given by,

$$\Delta r = \frac{2l_p{}^2}{\alpha_e R_s}\Delta n$$

$\Delta r = 0.3406 \times 10^{-16} m$

In conclusion the discrepancy is due to the anomalous magnetic dipole moment and it will never go away.

16 THE GALAXY ROTATION PROBLEM

The Big problem: Is dark matter responsible for differences in observed and theoretical speed of stars revolving around the centre of galaxies, or is it something else?

Why would we want to modify Einstein's outstanding intellectual achievement?

a) Newtonian and Einstein gravity cannot describe the motion of the outermost stars and gas in galaxies correctly.

b) If dark matter is not detected and does not exist, then Einstein's and Newton's gravity theories must be modified.

Since the 1970s and early 1980s, a growing amount of observational data has been accumulating that shows that Newtonian and Einstein gravity cannot describe the motion of the outermost stars and gas in galaxies correctly, if only their visible mass is accounted for in the gravitational field equations.

To save Einstein's and Newton's theories, many physicists and astronomers have postulated that there must exist a large amount of "dark matter" in galaxies and also clusters of galaxies that could strengthen the pull of gravity and lead to an agreement of the theories with the data. This invisible and undetected matter removes any need to modify Newton's and Einstein's gravitational theories. Invoking dark matter is a less radical, less scary alternative for most physicists than inventing a new theory of gravity.

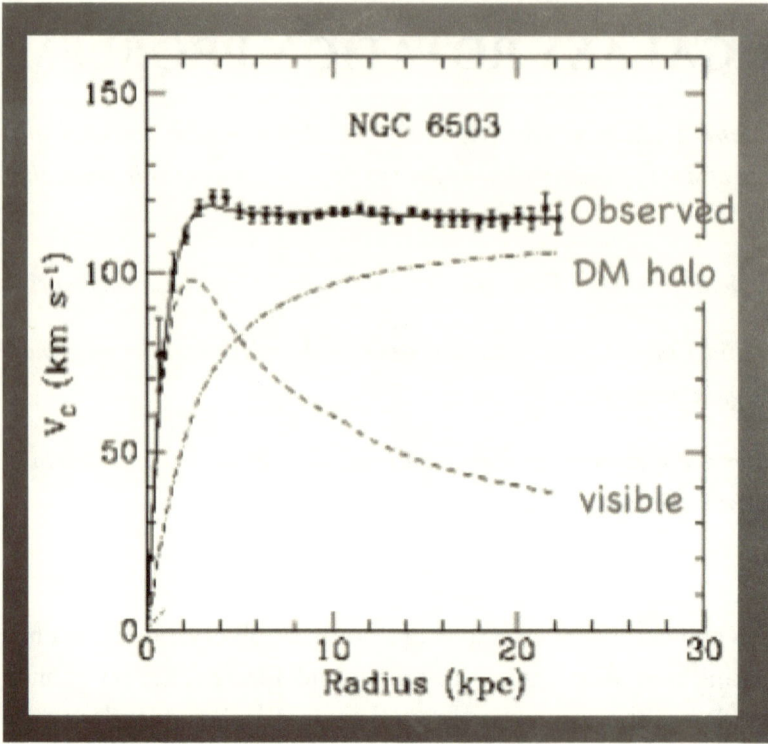

Fig. Galaxy data that show that Newtonian and Einstein gravity do not fit the observed speed of stars in orbits inside a galaxy such as NGC 6503

If dark matter is not detected and does not exist, then Einstein's and Newton's gravity theories must be modified. Can this be done successfully? Yes! My modified gravity (MOG) can explain the astrophysical, astronomical and cosmological data without dark matter.

(a)The Modified Force Law

Let the force of gravity be simplified as,

$$F = p\sqrt{\frac{2\pi \alpha g}{\lambda}}$$

(1)

Where p=mc, is the relativistic momentum of a particle of mass m. α is the coupling constant which determines the strength of a force in any interaction at the range determined by the Compton wavelength λ. $g = \dfrac{GM}{R^2}$ is the acceleration due to gravity for a point particle at a distance R from a star of mass M.

The above given law was used in chapter 10 and chapter 11 and it was expressed as, $F = \dfrac{me}{R}\sqrt{\dfrac{GM\Box}{4\pi\hbar\varepsilon_o}}$

Where $\omega=2\pi f$ is the angular frequency of the graviton-photon oscillations and e is the charge on an electron. In a limit of $\omega = \dfrac{GM\sqrt{4\pi\hbar\varepsilon_o}}{R^2}\left(\dfrac{1}{e^2}\right) = \dfrac{g_N}{\alpha_e}$, *where g_N is the usual Newtonian acceleration due to gravity and α_e is the fine structure constant, the above new force law reduces to the Newtonian law of universal gravitation.*

It must be noted that the above force law Eqn1 reduces to the Newtonian law of gravity $\dfrac{GMr}{R^2}$, only when the following two conditions are met:-

Condition1. The wavelength is equal to the circumference of the circle swept out by the orbiting mass m around M, that is $\lambda=2\pi R$.

Condition 2. When the coupling constant is half the Newtonian deflection angle θ for a light ray under the influence of a high gravitational field at the sun's surface ($\alpha =\theta/2=Rs/2R$, where Rs is the Schwarzichilds radius)

When the two conditions given above are met, then Eqn1 will become the Newton's law of gravity.

(b)The Tully-Fisher Relation

One of the best fit predictions of MOND is a single universal Tully-Fisher relation.

" The relation between asymptotic velocity and the mass of the galaxy is an absolute one" (Milgrom 1983). This is given by, $V^4 = a_oGM$, where $a_0 = 1.2 \times 10^{-10} ms^{-2}$. In this chapter an equation similar to the Tully-Fisher relation is deduced from (1) as given below,

For circular orbits about a mass M, we have the centripetal force equal to Eqn1 as,

$$\frac{mV^2}{R} = p\sqrt{\frac{2\pi ag}{\lambda}} = \frac{mc}{R}\sqrt{\frac{2\pi aGM}{\lambda}}$$

This gives an asymptotically rotation velocity independent of R:

$$V^4 = \left(\frac{2\pi ac^2}{\lambda}\right)GM = a_0 GM$$

$$(2)$$

It is this behavior that gives rise to *asymptotically flat rotation curves* and the Tully-Fisher relation (Tully & Fisher 1977) without invoking dark matter.

Comparing (2) to the Tully-Fisher relation, we determine the acceleration limit as,

$$a_0 = \frac{2\pi ac^2}{\lambda}$$

From which the coupling constant takes on values of,

$$\alpha = \frac{a_0\left(\frac{\hbar}{mc}\right)}{2\pi c^2}$$

This result implies that, if the milgrom acceleration was really a constant of $a_0 = 1.2 \times 10^{-10} ms^{-2}$ and also the fine structure constant was the coupling constant of $1/137$, then it will be true that the wavelength or the range of the interaction will be given by exactly ,$\lambda = 3.44 \times 10^{25} m$ which is within the acceptable size of the galaxies. This could help us connect quantum mechanics with gravity at small accelerations.

We anticipate that MOG will modify how stars collapse and the nature of black holes.We know that a supermassive object with mass $\sim 3 \times 10^6$ MSUN is at the center of our Galaxy (MILKY WAY). We are not able to determine yet whether the object is a GR black hole with a horizon. Perhaps future telescopes and space missions will be able to get close enough to the supermassive object to tell whether it is a black hole in spacetime or some other kind of object. However, as distant observers, we can never see a black hole event horizon form! The formation of the event horizon occurs in the infinite future, so we cannot actually ever see a black hole event horizon form as a star collapses.

(c)Modification of the Newtonian Dynamics (MOND)

From equation (c) Chapter13 we showed that the gravitational acceleration of a particle is related to the kinetic energy W and magnetic flux Φ by the following equation,

$$a_g = \frac{Wq}{\Phi h \varepsilon_0}$$

The above equation implies that, keeping the energy a constant, the acceleration of a particle increases with a fall in the magnetic flux and vice versa is true. Thus the acceleration of a particle is to a great extent affected by the magnetic field.

From the above given relationship we deduce a modification of the Newtonian dynamics in a limit of small accelerations similar to the Milgrom hypothesis (Astrophys. J. 270, 365-1983) but with an electric force as a cause of these small accelerations and we deduce the value of this acceleration in relation to the movement of electrons in the hydrogen atom. Using the same analysis we also deduce the value of the total mass of the universe.

If we let the energy or the kinetic energy, the work done to move the particle around a body of mass M (a galaxy) in the magnetic field created by M, in a process of magnetic induction be given by, $W = mv^2$, where v is the velocity of the particle and m is the mass of the particle.Also if we let the magnetic flux at any point in space at a distance R from M be $\Phi = 4\pi R^2 B$, then the acceleration is simplified in this way,

$$a_g = \frac{mv^2q}{4\pi R^2 Bh\varepsilon_o}$$

This can then be re-written to resemble the Milogram formula in this way,

We know the magnetic field B is related to the velocity and electric field E by, Bv=E, then the acceleration of a particle is,

$$a_g = m\left(\frac{qa}{4\pi\varepsilon_o Ehv}\right)a$$

Where, $a = \frac{v^2}{R}$. This then reduces to,

$$a_g = \left(\frac{q^2 a}{4\pi\varepsilon_o hv\left(\frac{Eq}{m}\right)}\right)a$$

This then gives,

$$a_g = \left(\frac{a}{\frac{1}{a_e}\left(\frac{Eq}{m}\right)}\right)a$$

From the above given given equation, $\alpha_e = \frac{q^2}{4\pi\varepsilon_o hv}$ is the fine structure constant. Eq is the electric force on the particle of charge q. Then the acceleration of a particle of mass m in the electric field will be given by, $a_1 = \frac{Eq}{m}$.

Now relating this to the Milogram Hypothesis of $a_g = \left(\frac{a}{a_o}\right)a$, where $a_o \sim 1.2 \times 10^{-10} m/s^2$, we have

$$\left(\frac{a}{\frac{1}{\alpha_e}a_1}\right)a = \left(\frac{a}{a_o}\right)a$$

Therefore in a limit where v=c (the speed of light) and q=e (the charge on an electron), we have

$$\alpha_e = \frac{e^2}{4\pi\varepsilon_o hc} = 1/137$$

This implies that, the acceleration of a particle in an electric field of a body of mass M will be given as

$$a_1 = \alpha_e a_o = 8.759 \times 10^{-13} m/s^2$$

$$(d)$$

The above given value implies that, in the presence of the electric and magnetic field the Milogram acceleration constant must be corrected to fit the data very well.

Then the modified Newtonian law of gravity is written as,

$$F = \frac{GMm}{\left(\dfrac{\alpha_e a}{a_1}\right)R^2}$$

In a limit $a_1 = a$ we don't recover the Newtonian law except for $\alpha_e = 1$, which proves to be difficult. This therefore requires us to rethink gravity with a new gravitational constant that could read as $G_B = \dfrac{G}{\alpha_e} = 9.138 \times 10^{-09} Nm^2/kg^2$

(d)Calculation Of The Total Mass Of The Galaxy From Equation (d)

Assuming the ratio of the accelerations to be equal to the fine structure constant as,

$$\frac{g_b}{g_a} = \alpha_e$$

$$(e)$$

Where,

$$g_a = \frac{8\pi\varepsilon_o hc^5}{e^2 GM}$$

And,

$$g_b = \frac{c^4}{GM}$$

Relating these accelerations to the ones given in equation d we should be able to deduce the mass of the galaxy as,

$$g_a = \frac{8\pi\varepsilon_o hc^5}{e^2 GM} = a_o = 1.2 \times 10^{-10} m/s^2$$

From which the total galaxy mass is given as,

$$e^2 M = 7.122 \times 10^{18} kg C^2$$

For $e = 1.602 \times 10^{-19} C$, the total mass is given as,

$$M = 5.55 \times 10^{56} kg$$

This is almost the total mass of the Universe. *Note: since the total mass of ordinary matter in the universe is known to be* $M = 1.5 \times 10^{53} kg$, *then using the above given equation, the charge required to give this value is* $e = 6.891 \times 10^{-18} C$

But with,

$$g_b = \frac{c^4}{GM} = \alpha_e a_o = 8.759 \times 10^{-13} m/s^2$$

We have,

$$M = 1.387 \times 10^{56} kg$$

Because this is not the same mass obtained with the use of g_a, we then rewrite the equation of g_b to give the same value of mass as,

$$g_b = \frac{nc^4}{GM}$$

Where, n is the number taking on values from, n=2,4,16,..............

In conclusion, we have not only modified the Newtonian dynamics as a requirement to account for darkmatter but we have just deduced the total mass of the universe. This implies that we are near or probably we have just completed the theory of quantum gravity.

17 THE IRREDUCIBLE ANOMALY IN THE OBSERVATIONS OF THE DEFLECTION OF LIGHT BY THE SUN

The big problem: Why Does the Eddington Value for the Deflection of Light Differ from the Freundlich Value in General Relativity?

It has long been suspected that the deflection of light in the vicinity of the sun exceeds the general relativistic predicted value of 1.75". An example of this, is the Erwin Finlay Freundlich 1929 solar eclipse expedition which produced a value of 2.24" larger than the general relativistic value. It is expected that once the reason for the deviation in the deflection angle has been found, it will disprove Einstein's imaginations for the curvature of space time. Although research into this field is scarce, we have managed through theoretical means under the assumption of a modified spherically symmetric solution to the Einstein field equation to prove E.F. Freundlich right. It is theorized that the bending of light near the sun is a function of the strength of the force (coupling constant) near the sun and the increasing distance from the sun's surface (in terms of the Schwarzschild radius).

It's almost hundred years since Sir Arthur Eddington experimentally proved Einstein's general relativity theory right. Since then, there has never been any competing theory that would prove Einstein wrong save for Loop quantum gravity and string theory. The fact that starlight is bent at the surface of the gravitating body by a deflection angle of 1.75" imposes a bound on the theoretical justification of gravity. Calculating an angle below or above 1.75" will be an upheaval in the founding blocks of physics. Erwin Finlay Freundlich was one of those people who stood out of the ordinary in 1929 when he published results with a larger angle of deflection than Eddington's.

An account on Freundlich 1929 expedition has been clearly given in Robert J.Trumpler and Klaus Hentschel papers as stated below;

"Among the various expeditions sent out to observe the total solar eclipse of May 9, 1929, that of the Potsdam Observatory (Einstein Stiftung) seems to be the only one which obtained photographs suitable for determining the light deflection in the Sun's gravitational field. Two instruments were used, but so far only the results of the larger one, a 28-foot horizontal camera combined with a coelostat, have been published. The three observers, Freundlich, von Klüber, and von Brunn, claim that these observations (four plates containing from seventeen to eighteen star images each) lead to a value of 2.24" for the deflection of a light ray grazing the Sun's edge; a figure that deviates considerably from the results of the 1922 eclipse, and which is in contradiction to Einstein's generalized theory of relativity".

The irreducible anomaly in the observations of the deflection of light by the sun has been known to exist since the birth of Einstein General relativity theory. For example, in a 1959 classical review by A.A.Mikhailov, it concludes that observations yield instead of a general relativistic prediction of 1.75arcsec at the limb of the sun the simple mean value of 2.03 ± 0.10 over the GR prediction.

The existence of a 2.24" deflection angle by Freundlich, Von Kluber and Von Brunn therefore implies a requirement for the modification of the general theory of relativity. Science has evolved in this simpler manner of modifications although there are some who cling to the old thoughts of "The earth is the center of the universe and Einstein is always right". I am not proving anyone wrong but I want you to believe that the general relativity theory that was put forward by Einstein is not the only 'there is' excellent description of the universe, there are other ways far better than GR as it was with the Newtonian Gravitational force replacement with a curvature of space time.

In this section I will prove Erwin Finlay Freundlich solar eclipse results right but from a theoretical perspective. We base our study on the bending of starlight past the surface of the sun, we establish the deflection angle at which this occurs starting from General relativity and beyond.

Einstein's theory proposes that gravity is not an actual force, but is instead a geometric distortion of spacetime not predicted by ordinary Newtonian physics. The more mass you have to produce the gravity in a body the more distortion you get, this distortion changes the trajectories of objects moving through space, and even the paths of light rays, as they pass close-by the massive body. Even so, this effect is very feeble for an object as massive as our own sun, so it takes enormous care to even detect that it is occurring.

General Relativity predicts how much of this bending of light you should see given the mass of the object. By formula, the Einstein General Relativity deflection angle is given by,

$$\theta_{GR} = \frac{4GM_\odot}{c^2 R_\odot} = \frac{2R_s}{R_\odot} = 1.75 \text{arcsec}$$

Where, $M_\odot$ − mass of Sun$(1.989 \times 10^{30}\text{kg})$

$R_\odot$ − Radius of Sun$(6.957 \times 10^8 \text{m})$ and $R_s = \frac{2GM_\odot}{c^2}$ is the Schwarzischild radius. The above given value doesn't come by surprise and it is the work of genius to find out why Newton, Einstein and others got different values. Below we provide a solution to the anomaly in the observation of light deflection at the sun's surface as given below.

For a test particle or an observer falling freely from infinity to a distance r_o from the gravitating body, the modified spherically symmetric solution to the Einstein field equation (see Eqn 4 in chap1) will be given by;

$$(\theta R)^2 = \frac{1}{\alpha^{1/2}}\left(\frac{R_s r_o}{2}\right)$$

Where, $R_s = \frac{2GM}{c^2}$ is the Schwarzschild radius of a gravitating body, α is the coupling constant and θ is the angle of deflection of a light ray past a gravitating body. This angle was never introduced in the previous chapter but it has been introduced in here for purposes supporting this research. In what follows, we use the above equation by subsitituting in the values of the dimensionless physical constants for the theory in question to obtain values of the deflection angle for each theory. This analysis will help us recover new theories based on the coupling constant and then draw conclusions.

Let us start with the **Newton's theory of gravitation**. To recover the Newtonian deflection angle at the suns limb, $r_0 = R_s$, we set the dimensionless physical constant to be $\alpha = 0.25$. This then gives the Newtonian value as,

$$\theta_1 = \frac{R_s}{R_\odot} = 0.875 \text{arcsec}$$

The **Einstein General relativistic** value can be got in the same way but this time with the dimensionless physical constant given by $\alpha = 0.0156$. This then gives the GR value as,

$$\theta_2 = \frac{2R_s}{R_\odot} = 1.75 \text{arcsec}$$

The **Freundlich deflection angle** might have taken a different twist than with Eddington 1.75arcsec result, which we are yet to find out and which is the reason for this expedition. Takin, $r_0 = R_s$ and $= 5.8208 \times 10^{-3}$, we deduce the deflection angle given by the formula,

$$\theta_3 = \frac{2.56R_s}{R_\odot} = 2.24 \text{arcsec}$$

Lastly when the dimensionless constant is the **fine structure** constant, $\alpha_e = \frac{1}{137}$ (note that, we have not used the gravitational coupling constant here simply because light as a photon is masssles). We get the following deflection angle,

$$\theta_4 = \frac{2.426R_s}{R_\odot} = 2.12 \text{arcsec}$$

Our first result from the above calculations is that; the mean of the deflection angles from the four observations gives the exact deflection angle that was calculated and observed by Eddington in General relativity as,

$$\frac{\sum_{n=1}^{4}\theta_n}{4} = \frac{0.875 + 1.75 + 2.24 + 2.12}{4} = 1.75 arcsec$$

The difference in the observed deflection angle is due to variations in the coupling constant or dimensionless physical constant from the modified metric which to a great extent differs from the Rindler, Schwarzschild, Reissner-Nordstrom, Kerr-Newman and Friedman-Lemaitre metrics. The fact that the mean of the four observations for the deflection of light given above reproduces the GR value of 1.75arcsec imposes a general bound on the dimensionless physical constants which is one of the unsolved problems in physics. Keeping other factors constant, the sum of four values of the coupling constants in any observation must not exceed the following value,

$$\alpha_1 + \alpha_2 + \alpha_3 + \alpha_4 = \frac{1}{256} = 3.90625 \times 10^{-3}$$

The model given above still assumes that general relativity is the correct theory of gravity on cosmological scales. If this was not correct, then the mean will have produced a different value of the deflection angle. But the fact that the mean reproduced the GR value makes it the correct theory of gravity from all the theories involved. Therefore our part is to determine the kind of theory behind the **Freundlich deflection angle.** And once this is found out then there will be transparence in the final theory, unified theory and the theory of everything.

18 THE EXTRA DIMENSION PROBLEM

The big Problem: (i) Is it true that at every spatial dimension, there exists new physics and that it is the work of Physicists to find out? What is the method or procedure through which new physics can be found? Does this require extra dimensions?

(ii) Does nature have more than four space-time dimensions? If so, what is their size? Are dimensions a fundamental property of the universe or an emergent result of other physical laws? Can we experimentally observe evidence of higher spatial dimensions?

(iii) Can the singularities that plague the General theory of Relativity be resolved in any quantum theory of Gravity?

History tells us that if we hit upon some obstacle, even if it looks like a pure formality or just a technical complication, it should be carefully scrutinized. Nature might be telling us something, and we should find out what it is (G. t Hooft, 1997).

In physics, one of the ultimate goals is to unify the fundamental forces of nature. Today physicists have been able to unify three of the four known fundamental forces (the electromagnetic, the strong and the weak nuclear forces in a single quantum field theory-the standard model). The fourth fundamental force, gravity, on the other hand is described by the general theory of relativity. Because the other fundamental interactions are quantized, it therefore seems natural that in a grand unified theory, a theory of all the fundamental forces, gravity is quantized as well into perhaps Quantum gravity.

A theory of quantum gravity is needed to describe things that are very small but also very heavy, like black holes or the early universe. However, the development of a quantum theory of gravity seems difficult on grounds that, in general relativity all physical qualities have definite values, whereas in quantum mechanics they do not as shown in Heisenberg's uncertainty principle.

The problems in General Relativity arise from trying to deal with a universe that is zero in size (infinite densities). But quantum mechanics suggests that there may be no such thing in nature as a point in space-time, implying that space-time is always smeared out, occupying some minimum region. The minimum smeared-out volume of space-time is a profound property in any quantized theory of gravity and such an outcome lies in a widespread expectation that singularities will be resolved in a quantum theory of gravity.

However, Prof Brian Dolan at the Department of Theoretical Physics, NUI Maynooth, is quick to point out that there is not yet any set agreement on what a theory of quantum gravity should look like, or even on the exact problem it is trying to solve."There is no accepted theory of quantum gravity," he says. "There are currently a number of contenders, and by far the most popular is superstring theory. Many physicists find superstring theory compelling due to its internal elegance, but despite decades of intense research it has not produced a single experimentally testable result." He suspects that trying to unite general relativity and quantum mechanics may be the wrong way to go, and that any future breakthrough may come from a completely unexpected direction; perhaps from some young mind with a fresh perspective.

This chapter employs new idea towards the development of a quantum theory of gravity in a bid to solve the following unsolved problems in physics;

(i) Is it true that at every spatial dimension, there exists new physics and that it is the work of Physicists to find out? What is the method or procedure through which new physics can be found? Does this require extra dimensions?

(ii) Does nature have more than four space-time dimensions? If so, what is their size? Are dimensions a fundamental property of the universe or an emergent result of other physical laws? Can we experimentally observe evidence of higher spatial dimensions?

(iii) Can the singularities that plague the General theory of Relativity be resolved in any quantum theory of Gravity?

The Standard Model is inconsistent with that of general relativity, to the point that one or both theories break down under certain conditions (for example within known spacetime singularities like the Big Bang and the centers of black holes beyond the event horizon).

The appearance of singularities in any physical theory is an indication that something is wrong and that there is a need for new physics. Singularities can be avoided in GR and any field theory through the introduction of an efficient regularization procedure as this book directs.

Regularization is a method of modifying observables which have singularities in order to make them finite by the introduction of a suitable parameter called regulator. The regulator, also known as a "cutoff", models our lack of knowledge about physics at unobserved scales (e.g. scales of small size or large energy levels). **It compensates for the possibility that "new physics" (beyond the SM) may be discovered at those scales which the present theory is unable to model,** while enabling the current theory to give accurate predictions as an "effective theory" within its intended scale of use.

The need for regularization terms in any quantum field theory of quantum gravity is a major motivation for Physics beyond the standard model. Infinities of the non-gravitational forces in QFT can be controlled via renormalization only but additional regularization and hence new physics is required uniquely for gravity. The regularizers model, and work around, the breakdown of QFT at small scales and thus show clearly the need for some other theory to come into play beyond QFT at these scales. A. Zee (Quantum Field Theory in a Nutshell, 2003) considers this to be a benefit of the regularization framework, theories can work well in their intended domains but also contain information about their own limitations and point clearly to where new physics is needed.

Therefore the main objective of this section is to discover new physics at those scales (or extra dimensions) which the General relativity theory and Quantum mechanics is unable to model. The section also sets out to prove that due to quantum gravitational effects, there is a minimum distance beyond which the force of gravity no longer continues to increase (operate) as the distance between the masses become shorter.

General Theory

During the years, strong evidence has appeared that the acceleration of any physical object cannot be arbitrarily large, but it should be superiorly limited. For example in string theory, it was derived that string acceleration must be less than some critical value, determined by the string tension and its mass. From the classical point of view (as Wheeler suggested), if we consider an extended object in **rotating motion**, we have the acceleration $a = v^2/R$ and it follows that a, must be at least limited by c^2/R. However to differ from the classical Newtonian mechanics and Einstein's General relativity theory we introduce a regulator "Cutoff" $\alpha_g{}^n$,where α_g is the gravitational coupling constant, R is the distance between two masses and n is a positive number (**extra dimension number**), then the acceleration must be limited by $a = \frac{c^2}{2R}\alpha_g{}^n$ (i), (Assuming a diameter of 2R).

Thus to avoid the infinity but while retaining the point nature of the particle would be to postulate a small additional dimension **n** over which the particle could 'spread out' rather than over 3D space.

For example, in the Unruh temperature we can only and only deduce both the Hawking temperature and maximal temperature (Sakharov Temperature) under the assumption of the existence of a maximal acceleration given in formula (i) above as,

The Unruh temperature is given as,

$$T = \frac{\hbar a}{2\pi c k}$$

Since the acceleration is known from (i) above, then the temperature will reduce to,

$$T = \frac{\hbar c}{4\pi R k}\alpha_g{}^n$$

For a Schwarzschild Black hole of radius $R = \frac{2GM}{c^2}$, the temperature reduces to

$$T = \frac{\hbar c^3}{8\pi GMk}\alpha_g{}^n$$

Since the gravitational coupling constant has a formula $\alpha_g = \frac{GM^2}{\hbar c}$, taking values of n=0,1,2,............,N. Then the Hawking temperature will become a result of n=0 extra spatial dimensions as,

$$T = \frac{\hbar c^3}{8\pi GMk}$$

.

Also the maximum temperature (Sakharov temperature) is deduced at n=1/2 as ,

$$T = \frac{1}{8\pi k}\left(\frac{c^5\hbar}{G}\right)^{1/2}$$

Therefore the temperature of a black hole increases as a black hole loses mass in Hawking Black hole evaporations. The analysis given above is a clear indication that the temperature doesn't increase exponentially as it has been known from Hawking's original proposals, there is a maximum temperature, a limit on temperature that screens (resolves) the classical singularity. It is therefore true that the radiation spectrum contains all Standard Model particles, which are emitted on our brane, as well as gravitons, which are also emitted into the extra dimensions. It is expected that most of the initial energy is emitted during this phase in Standard Model particles. Therefore we recommend the applications of a factor $\alpha_g{}^n$ in situations involving the examination and experimentation of quantum gravitational phenomenon. We shall see in the coming chapters that such a factor when used in loop quantum cosmology it reproduces both the results of loop quantum gravity and string theory.

The idea of including extra dimensions, to achieve the goal of unifying physics, is not a new one. Already the year before Einstein in 1915 introduced his theory of general relativity; Gunnar Nordstrom suggested a unification of gravity and electromagnetism with the introduction of a fifth dimension. These forces were the two only forms of interaction known at that time. But this idea was forgotten for some time with the eruption of the First World War. But in April 1919 Theodor Kaluza introduced independently, in a letter to Einstein, a fifth dimension in an attempt to unify Einstein's theory of gravity and Maxwell's theory of light. Oskar Klein (1926) contributed, in this quest, with his assumption that the extra dimension was compactified. The Kaluza-Klein theory was a fact. This theory includes an extra space dimension that is rolled up into a tiny circle, i.e. compactified. And in this five dimensional theory, there is only one underlying force, gravity. But in the four-dimensional spacetime observed at great distances, it appears to be three kinds of forces, among these a gravitational and an electromagnetic force. This topic was initially a popular topic for research, but lost much of its interest with the introduction of quantum mechanics.

In recent years the topic of extra dimensions has experienced a renewed interest. This renewed interest is also due to the exciting possibility of observing new and spectacular physical phenomena at far lower energy scales than otherwise. Even at energies available in the not so distant future, these phenomena could appear. Among these is the creation of higher dimensional semi-classical microscopic black holes. The possibility of observing these objects, is viewed as an opportunity to perhaps discover new intriguing physics.

Therefore from (i) using Einstein's equivalence principle we get the minimum distance beyond which the force of gravity no longer continues to increase as;

$$R = \frac{R_s}{\alpha_g{}^n}$$

(ii).

Where $R_s = \frac{2GM}{c^2}$ is the Schwarzschild radius. We therefore conclude that;

(i) At n=0 extra spatial dimension, we have a physical theory of General relativity at a length scale of $R = R_s = \dfrac{2GM}{c^2}$ - the Schwarzschild radius.

(ii) At n=1/2 extra dimension, we have the quantum theory of gravity (New physics) at the Planck length scale $l_p = \sqrt{\dfrac{\hbar G}{c^3}}$

(iii) At n=1 extra dimension, we have the theory of Quantum mechanics at the Compton wavelength scale of $\lambda = \dfrac{h}{mc}$.

(iv) Lastly at n=2 we have new physics at a length scale $R = \dfrac{h^2}{GMm^2}$ and the journey continues.

According to the Standard Model of particle physics, the world is governed by four fundamental forces: gravity, electromagnetism, and the weak and strong nuclear forces. Although things act a bit "spooky" down on the quantum level, science has managed to generally describe all of these forces at both the macro and quantum scales – except gravity.

Gravity is the weakest of the fundamental forces, and it's been suggested that this is because some gravitons (the hypothetical particles) that carry the gravitational force tend to escape into extra dimensions. We're simply too big to travel through or even notice these other dimensions.

So, to study whether these extra dimensions are lurking in extremely tiny spaces, the researchers from Osaka, Kyushu and Nagoya Universities set out to test gravity on the sub nanometer scale. To do so, they used the world's highest intensity neutron beam, which is housed at the Japan Proton Accelerator Research Complex (J-PARC).

The team found that the results matched predictions based on the known laws of physics, which indicates that Newton's law still applies as expected down to a scale of less than 0.1 nanometers. No unexplained force ie, another dimension is acting on these particles at this scale.

That doesn't mean those extra dimensions aren't there, just that they may be hiding at even smaller scales still. The researchers are currently working to further improve the sensitivity of the equipment, which might help them probe those tiny spaces.

In a completely different context, an international team of researchers led by Professor Immanuel Bloch (LMU/MPQ) and Professor Oded Zilberberg (ETH Zürich) has now demonstrated a way to observe physical phenomena proposed to exist in higher-dimensional systems in analogous real-world experiments. Using ultracold atoms trapped in a periodically modulated two-dimensional superlattice potential, the scientists could observe a dynamical version of a novel type of quantum Hall effect that is predicted to occur in four-dimensional systems. (Nature, 4 January 2018)

"Physically, we don't have a 4D spatial system, but we can access 4D quantum Hall physics using this lower-dimensional system because the higher-dimensional system is coded in the complexity of the structure," a researcher with the US-based team, Mikael Rechtsman from Penn State University, told Ryan F. Mandelbaum at Gizmodo. "Maybe we can come up with new physics in the higher dimension and then design devices that take advantage the higher-dimensional physics in lower dimensions."

The above statements can be summed up in the following simplest model;

Let the Gravitational force between two identical particles be related to the magnetic force between them and similarly let the electric force between two particles be related to the magnetic force as;

Gravitational force $(\frac{Gm^2}{R^2})$ = magnetic force (Bec) $\times \alpha_g{}^n$ (iii)

and

Electric force $(\frac{e^2}{4\pi\varepsilon R^2})$ = magnetic force (Bec) $\times \alpha_e{}^n$ (iv)

Where α_e is the electromagnetic coupling constant- Fine structure constant

The magnetic flux, represented by the symbol $\mathbf{\Phi}$, threading some contour or loop is defined as the magnetic field $\mathbf{B}$ multiplied by the loop area, $A = \pi R^2$, i.e. $\mathbf{\Phi} = \mathbf{B} \cdot \mathbf{A}$. Obviously, both $\mathbf{B}$ and $\mathbf{A}$ can be arbitrary and so is $\mathbf{\Phi}$. The inverse of the flux quantum, $1/\Phi_0$, is called the **Josephson constant**, and is denoted K_J.

However, if one deals with the superconducting loop or a hole in a bulk superconductor, it turns out that the magnetic flux threading such a hole/loop is quantized. Therefore the magnetic flux quantum from (iii) and (iv) will be given by,

$$\Phi_G = \pi G \, m^2 \, \alpha_g^{\,n} / ec$$

$$\Phi_E = e/4\varepsilon c \, \alpha_e^{\,n}$$

Such that at n=0 extra dimension,

$$\Phi_G = \pi G \, m^2 / ec$$

$$\Phi_E = e/4\varepsilon c$$

The above given values represent the classical flux at 3D spatial dimensions.

At n=1/2 extra dimension,

$$\Phi_G = \frac{\pi m}{e}\left(\frac{Gh}{c}\right)^{1/2}$$

$$\Phi_E \ = \ \left(\frac{\pi h}{4\varepsilon c}\right)^{1/2}$$

These represent the quantum theory of Gravity.

At n=1 extra dimension,

$$\Phi_G = \pi\hbar/e$$

$$\Phi_E = \pi\hbar/e$$

These represent the magnetic flux quantum at the quantum scale. Also at n=1 the magnetic flux value is the same in both equations, meaning that the gravitational force becomes analogous to the electromagnetic force at n=1.

In other words, just as a 3D object casts a 2D shadow, scientists have managed to observe a 3D shadow potentially cast by a 4D object – even if we can't actually see the 4D object itself. That could unlock some new findings in the very fundamentals of science.

PART IV BRIEF LESSONS ON PHYSICS

19 THE MOST BEAUTIFUL OF THEORIES

Physicists have argued out that the more elegant and symmetrical the theory is, the more it is beautiful. The elegancy of any physical theory is suspected at a level to which it holds well with other theories , that is ,the capability of the theory to conform with the well known laws of nature at all levels.

In this section we examine the mechanism through which quantum mechanics becomes comparable with gravity and the scale at which this occurs. At the Planck scale all interactions (the weak interaction, strong interaction and electromagnetism) are assumed to merge into a single interaction that alone occurs at very high energies of about 1TeV. The equations that do describe this phenomenon are not yet found and therefore requires one's deep effort to capture the reality of this entire puzzle.

To capture interest in these interactions we need to know first, their strength and second the range in which they occur. The strength defines the coupling constants and the range defines the attractions, on the other hand the coupling constant determines the strength of any interaction and therefore is a number in a sense that it is a dimensionless constant. A coupling constant is a very important quantity in dynamics, for example, in the motion of a large lump of magnetized iron, the magnetic forces are more important than the gravitational forces because of the relative magnitudes of the coupling constants.

The standard model is a theory of three fundamental forces - electromagnetism, weak interactions and strong interactions; however, these three forces are not tied together Howard Georgi and Sheldon Glashow discovered that the Standard Model particles can arise from a single interaction, known as a grand unified theory. Grand unified theories predict relationships between otherwise unrelated constants of nature in the Standard Model. Gauge coupling unification is the prediction from grand unified theories for the relative strengths of the electromagnetic, weak and strong forces and this prediction was verified at LEP in 1991 for supersymmetric theories.

In particle physics, supersymmetry (often abbreviated SUSY) is a novel symmetry that relates elementary particles of one spin to another particle that differs by half a unit of spin and are known as superpartners. Since the particles of the Standard Model do not have this property, supersymmetry must be a broken symmetry allowing the 'sparticles to be heavy.

One of the main motivations for SUSY comes from the quadratically divergent contributions to the Higgs mass squared. The quantum mechanical interactions of the Higgs boson causes a large renormalization of the Higgs mass and unless there is an accidental cancellation, the natural size of the Higgs mass is the highest scale possible. This problem is known as the hierarchy problem Supersymmetry reduces the size of the quantum corrections by having automatic cancellations between fermionic and bosonic Higgs interactions. If supersymmetry is restored at the weak scale, then the Higgs mass is related to supersymmetry breaking which can be induced from small non-perturbative effects explaining the vastly different scales in the weak interactions and gravitational interactions. The failure of experiments to discover either supersymmetric partners or extra spatial dimensions, as of 2006 has encouraged loop quantum gravity researchers.

(a)The determination of the strength of the forces

We assume a model that explains everything on the length scales, the best scale so far we are familiar with is the Planck length scale, however in this model we don't associate ourselves in knowing this scale and therefore develop new scales that alone are combined together to lead to some observable phenomenon describing the forces involved in the interactions. The equation describing the model is developed and given by;

$$(v^2/c^2 + n^2\beta_{Qo}) = 8\pi\beta_{gEo} \,(1)$$

Where β_{Qo} is a length ratio given by l_Q/l_o, in this case $l_Q = \hbar c/W$, $\hbar$ is Dirac constant, c is the speed of light and W is the energy. Also $\beta_{gEo} = l_{gE}/ l_o$ where $l_{gE} = 8\pi GM_{gE}^2/W$, G is the universal gravitational constant , M_{gE} is the mass of a particle in the combined fields given by P_{gE} /c where $P_{gE} = Gm^2 ke^2/R^2 c^2$

,is the momentum for an elementary particle of mass m and an elementary charge e , k is the coulomb constant and R is the distance between any two particles. The equation here addresses the problems in form of length scales simply because it is at these scales that quantum mechanics seem to be comparable to gravity. The momentum P_{gE} is a momentum of a particle experiencing the strength of the electromagnetic fields and gravity. The strength is determined by a very small coupling constant as we shall see later. The smaller the distance between elementary particles, the higher the momentum and vice versa is true.

The exchange of photons between an electron and a proton in an atom is explained by Quantum Electrodynamics (QED), with a coupling constant determining the strength of the electromagnetic force. The equation of the interaction responsible for QED on the length scale, which is the Compton length, is given by the equation

$$\Sigma \psi^2 \text{fiti} = 2\pi \beta_{RCE} \text{ (2)}$$

The expression $\Sigma \psi^2$fiti is the force changer where $\psi_{fi} = \text{fi}R^2/ke^2$ and $t = \{\text{fi}^2 ke^2/R^2\}/F_n^3$,

β_{RCE} ,remains a constant given by lc / lRE (lc is the Compton length $\hbar/mc$ and lRE=ke^2/mR^2c^2).On multiplying both sides of Eqn1 by a quantity $\Sigma \psi^2$fiti we obtain,

$$(v^2/c^2 + n^2 \beta_{Qo}) \, \Sigma \psi^2 \text{fiti} = 8\pi \beta_{gEo} \Sigma \psi^2 \text{fiti}$$

We then examine the condition for which β_{Qo} will be a maximum and minimum. It is found out from relativity that β_{Qo} is maximum when the lorentz factor $\gamma = (1 - v^2/c^2)^{-1/2}$ is very small that is,

$\gamma = 1/n\sqrt{} \, \beta_{Qo}$ or when the velocity $v = c \sqrt{(\Sigma \psi^2 \text{fiti} - n^2 \beta_{QO})}$

We hence obtain a general interaction equation as,

$$\Sigma f_y{}^3 \psi^2 fiti = 2\pi\beta RCE \Sigma Fn^3 / \xi\beta gEo \, , \, n= 1,2,3 \, (3)$$

The following conditions are then taken into account

1) For $lo = lx = mc^2/Fp$, $\beta gEo = \beta gEx = lgE/lx$.

2) For $lo = lc$, $\beta gEo = \beta gEQ = lgE/lc$, and

3) For $lo = ls = Gm/c^2$, $\beta gEo = \beta gEs = lgE/ls$, which gives

$$\Sigma f_y{}^3 \psi^2 fiti = F_1{}^3 + F_2{}^3 + F_3{}^3 = 2\pi\beta RCE \, (\, Fp^3 /8\pi\beta gEx + Fp^3 / 256\pi^3 \beta gEQ + 2FB^3 / \pi \, \beta gEo \,) \, (4)$$

Where $Fp = c^4/G$ is the Planck unit force and $FB^3 = m^2c^3/\hbar$ is the force required for strong and weak interactions to take place. Again setting a condition,

For for $lo = lz = ke^2/mc^2$, $\beta gEo = \beta gEz = lgE/lz$.

$$\Sigma f_y{}^3 \psi^2 fiti = F_4{}^3 = 2\pi\beta RCE \, (\, Fz^3 /32\pi^3 \beta gEx)$$

Where $Fz^3 = m^2 c^4/ke^2$,

Also for $lo = lN = \hbar^2 m^3 G^2/k^3 e^6$, $\beta gEo = \beta gEN = lgE/lN$, we obtain,

$$\Sigma f_y{}^3 \psi^2 fiti = F_5{}^3 = 2\pi\beta RCE \, (Fz^3/2\pi^3 \beta gEN) \, (5)$$

Measuring the value of the strong, weak and electromagnetic coupling constants gives us away through which we can determine supersymmetric levels. From supersymmetry and grand unification of elementary particles the couplings agree to 1%. The relationships of the sum of the cubes of the forces to each individual cube of the force, and that of the sum of the square of masses with each known mass squared casts much information about the masses and couplings of the supersymmetric particles as shown below, when Eqn4 is divided through respectively by the cubes of the forces F_1^3 , F_2^3 and F_3^3 the following equations are obtained,

$$\Sigma F_n^3/F_1^3 = 1 + 16\alpha_g^3 + 1/32\pi^2\alpha_g \ (6)$$

$$\Sigma F_n^3/F_2^3 = 1 + 32\pi^2\alpha_g + 512\pi^2\alpha_g^4 \ (7)$$

$$\Sigma F_n^3/F_3^3 = 1 + 1/6\alpha_g^3 + 1/512\pi^2\alpha_g^4 (8)$$

$$\Sigma F_n^3/F_4^3 = 1 + \beta^2(4\pi^2 + 1/8\alpha_g) + 64\pi^2\alpha_s^2\alpha_g \ (9)$$

Where $\beta = ke^2/Gm^2$ is the ratio of the fine structure constant α_s to the gravitational coupling constant α_g, given respectively as $\alpha_g = Gm^2/\hbar c$ and $\alpha_s = ke^2/\hbar c$.

Now equating $F_4 = F_5$, $F_5 = F_3$, $F_5 = F_1$ we obtain; m_1, m_2 , $m_3$and $m_4$respectively, Adding the squares of the masses we obtain,

$$\Sigma m_n^2 = m_1^2 + m_2^2 + m_3^2 + m_4^2 (10)$$

Which gives the sum per unit mass as,

$$\Sigma m_n^2/m_1^2 = 1 + 16\pi^2\alpha_s^4 + (8\pi/\alpha_s)^{\frac{1}{2}} + 4/(128\ \alpha_s^4)^{1/5} \ (11)$$

$$\Sigma m_n^2/m_2^2 = 1 + 1/16\pi^2\alpha_s^4 + (1/8\pi\ \alpha_s^9)^{\frac{1}{2}} + (1/4\pi^2)(128\ \alpha_s^{24})(12)$$

The equations generated so far give a basis for the nature and type of supersymmetry exhibited by a particle experiencing forces at both the Planck and grand unified scales. It is thus shown here that the electromagnetic coupling constant is a result of mathematically summing the squares of the masses generated and then dividing through by the square of the mass in the summation while the gravitational coupling constant is the result of summing the cubes of the forces and then dividing through by the cube of the force in the sum. This idea at its best is taken to be the basis for symmetric theories as we shall see in the results obtained.

(b)Results

(i)The unification of coupling calculations

At equal forces that is $F_1 = F_2 = F_3 = F_p$ the mass $M_p = (\hbar c/8\pi G)^{1/2} = 2.1765 \times 10^{-8}$ **kg**, is obtained which is the Planck mass for which the Schwarzschild radius is equal to the Compton length divided by π^1.When Eq4 is divided through by $F_1{}^6$ and $F_2{}^6$ we obtain equations of the form;

$$\sum F_n{}^3/F_1{}^6 = \Omega/F_p{}^3 \ (13)$$

$$\sum F_n{}^3/F_2{}^6 = \text{€}/F_p{}^3 (14)$$

Where, $\Omega = 4m^2/m_p{}^2 + 1/\pi + m^8/32\pi^2 m_p{}^8$ *and* $\text{€} = m^6/8\pi m_p{}^6 + 16\pi m^4/m_p{}^4 + 2m^{12}/\pi m_p{}^{12}$

The mass relations equations obtained above indicate the scale at which gravity may be strong and weak. Obtaining these results on the Planck force and mass scale is evidence for the existence of the theory of quantum gravity. The values Ω and € represent a series equation defined by increasing powers in the mass ratio (m/m_p). The mass **m** is assigned to any particle and the mass m_p is assigned to the Planck scale defining quantum gravity.

The unit of energy is $M_p c^2$; the unit of electric charge is $\sqrt{hc/k}$, where k is coulomb constant and so forth. On the other hand, one cannot form a pure number from these three physical constants. Thus one might hope that in a physical theory where $\hbar$, c, and G were all profoundly incorporated, all physical quantities could be expressed in natural units as pure numbers. Within its domain, this paper has achieved it for example, imagining that there were just two quark species with vanishing masses. Then from the two integers 3 (colors) and 2 (flavors), $\hbar$, and c (without mass parameters), the spectrum of hadrons with mass ratios and other properties close to those observed in reality, emerges by through calculation (Ω and €) as indicated from Eqn13 and Eq14 shown above. The overall unit of mass is indeterminate, but this ambiguity

---

1. http://en.wikipedia.org/wiki/Pi

has no significance within the theory itself. The results obtained show an ideal Planckian theory that alone does not contain any pure numbers as parameters.

Thus, for example, the value $m_e/m_p = 10^{-22}$ of the electron mass in Planck units is obtained from a dynamical calculation. This ideal might be overly ambitious, yet it seems reasonable to hope that significant constraints among physical observables will emerge from the inner requirements of a quantum theory which consistently incorporates gravity. The model therefore provides; first, the unification of couplings calculation. second, it points to a symmetry breaking scale remarkably close to the Planck scale (though apparently smaller by 10^{-2} to 10^{-3}), so there are pure numbers with much more 'reasonable' values than 10^{-22} to shoot for. Third, it shows quite concretely how very large scale factors can be controlled by modest ratios of coupling strength, due to the logarithmic nature of the running of couplings (so that 10^{-22} may not be so 'unreasonable' after all).

While the above result is based on the study of the strength of the gravitational force, we now look for ways in which we can examine the strength of the electromagnetic force depending on the mass. This is done by dividing the sum of the squares of the masses (Eqn10) by the fourth power of the individual masses hence,

$$\Sigma m_n^2 / m_2^4 = \omega / m_G^2 \ (15)$$

$$\Sigma m_n^2 / m_E^4 = \lambda / m_p^2. \ (16)$$

Where $\omega = 1 / \ 16\pi^4 \alpha_s^9 + 1 \ / 4\pi^2 \alpha_s^5 \ +1 / \ (512\pi^8 \ \ \alpha_s^{19} \ )^{1/2}$,
$\lambda = 128\pi^3 \alpha_s^6 + 8192\pi^5 \alpha_s^{10} + 128\sqrt{\pi^3} \alpha_s^{11} + 128(\pi^{15}\alpha_s^{26})^{1/5}$

$m_E = (1 / \ 8\pi \ Ke^2)(\hbar^3 c^3 / G)^{1/2}$ is the mass obtained when $F_4^3 = F_3^3$, and $M_G = (Ke^2 / G)^{1/2}$ is the mass obtained when the electromagnetic force is equal to the gravitational force.

It can now be theorized that the strength of the electromagnetic force is determined by Eqn15 and 16 at which a series power equation in the fine structure constant defined by ω and λ is a constant.

(ii)The length scales at which the masses predicted by the standard model survive

The mass of the Wand Z bosons (M_W, M_Z),Higgs particle (M_H) and the mass scale at the grand unification (M_{GUT}) are generated. We multiply a coupling constant μ with the force $F_3{}^3$, of which we equate to $F_4{}^3$ that is;

$$\mu \, F_3{}^3 = F_4{}^3$$

From which

$$\mu = R_B{}^2 / \, R_0{}^{2,}$$

Where R_0 is the length scale determined experimentally and R_B =(8 $\pi Gke^2/c^4)^{\frac{1}{2}}$=6.9101×10^{-36} m ,which is greater than the Planck length.

So the equation that produces the different masses at R_0 will be given by the square of the mass as,

$$M^2 = m_p{}^2/8 \, \pi\mu \, \alpha_s{}^2$$

Where $\alpha_s = 1/137$,is the electromagnetic coupling constant.

To obtain the masses, we need to find the length R_0 , theoretically we develop the lengths given by; 1.03741×10^{-39} m, 8.3182×10^{-54}m, 9.4334×10^{-54}m, and 1.2345×10^{-53}m.

Following the given lengths we respectively obtain the masses;

M_{GUT}=10^{16}GeV, M_W =80.18GeV , M_Z =90.82GeV, and M_H =119GeV respectively.

But at $R_B = R_o$, the mass $M_B = 6.661 \times 10^{19}$ GeV is obtained. And at $R_o = 2.529 \times 10^{-37}$ m, the Planck mass is obtained (that is $M = m_p$). Therefore it is found out that the W and Z boson particles survive in length of 10^{-54} m . The Higgs particle survives to a length greater than that of the W boson $\geq 10^{-53}$ m. And finally particles at the grand unified scale will survive at 10^{-39} m.

(iii) The big bang acceleration and proton decay

For proton decay the intensity P is used such that at Schwarzschild radius R and Planck mass scale m_p the life time of the proton as explained by SUSY is seen to agree so well with the

$$T(time) = \alpha^2 \, m_p^5 \, R \, / \, 4096 \, \pi^3 m k^4 \, \hbar$$

Such that at $mk = 7.96 \times 10^{-29}$ GeV, $T = 10^{35}$ yrs.

We have obtained the lifetime of protons and the mass of a particle produced during the decay process. The mass of the particle obtained is very small and can therefore be taken to be a neutrino.

The force F3 can be expressed in the form,

$$F_3 = a_3 (m_3^5 \, / 16\pi^2 m_p^2)^{1/3}$$

Where a_3 is the acceleration, this acceleration at a Planck scale will be given by

$$a_3 = (c^{11} \, / \, \hbar \, G^2 m)^{1/3} = 2.4772 \times 10^{52} \text{m/s}^2$$

This is quite a very large acceleration and therefore defined as the acceleration of particles during the early formation of the universe.

The results obtained describe super symmetry which is a theory required for the unification of everything we know about the physical world into a theory of everything. Significantly a larger enterprise of the theory is to produce a theory of quantum gravity which is required for the unification of general relativity with the standard model, which explains the other three basic forces in physics (electromagnetism, the strong interaction, and the weak interaction), and provides a palette of fundamental particles upon which all four forces act. Theoretically the results obtained (Eqn11and Eqn12) show a huge correction to the particles' masses, which without fine-tuning will make them much larger than they are in nature. The problem of the unification of the weak interactions, the strong interactions and electromagnetism is solved mathematically, through the comparisons of the cube of the forces in a ratio that generates the gravitational coupling constant power equation.

The Planck mass is the mass of a black hole whose Schwarzschild radius multiplied by π equals its Compton wavelength. The radius of such a black hole is roughly the Planck length, which is believed to be the length scale at which both general relativity and quantum mechanics simultaneously become important. In accordance with the results obtained it is seen that the Planck mass is the mass at which the four forces (F_1 , F_2 ,F_3 and F_p) are equal, the forces are then taken to be related to the origin of the universe simply because at those high energies that formed the dense soup of the universe the forces were equal and the masses probing the Planck mass scale that is black holes were produced, hence those four forces a significant in that they play a crucial role in the formation of black holes. The intensity P on the other hand explains a phenomenon that occurs at the cosmic scale, for example it explains the nature of Black holes and the age of the universe. The acceleration obtained is so large that it is the acceleration that the universe had at the instant after the big bang. Obtaining this acceleration is the possibility of studying the rate of expansion of the universe at large, the accelerating universe is therefore the observation that the universe appears to be expanding at an accelerated rate.

At the Planck scale the descriptions of subatomic particle interactions in terms of quantum field theory breaks down. Also at the same scale, the strength of gravity is expected to become comparable to the other forces, mathematically all the fundamental forces are unified at that scale. The results obtained explain both the weak and strong interactions that at a length between 10^{-37}m and10^{-35} the Planck scale is attained also at lengths10^{-39}m , the grand unified scale becomes relevant , but for lengths10^{-53}m and 10^{-54}m, the standard model holds on well. We have therefore attained a unification that increases from about 10^{-59}m (standard model) to10^{-35}m (quantum gravity).The paper there fore gives out the relationship between elementary particle physics and astrophysics at a large scale.

Basing on the results obtained, it is now clearly justified that gravity can be integrated with quantum mechanics at the Planck scale. And therefore the success of the "standard model" which includes both the electroweak theory and quantum chromodynamics can now be regarded as successful in providing accurate descriptions of the fundamental particles and their interactions.

20 QUANTA IN GENERAL

Basing our study on the electric currents generated whenever there is a changing magnetic field (B) and a changing electric field (E) in the electromagnetic wave we can construct a complete theory for the electromagnetic radiations. The theory is created using the symmetry between a long wire placed in the electromagnetic fields which induce vibrating electrons that carry current in the wire and the electromagnetic wave which constitute changing electric and magnetic fields that create vibrating photons in the wave. Therefore a wire is to a wave what a vibrating electron is to a vibrating photon in the wire and a wave respectively. The aim of the paper is to give a clear description of the theory of electromagnetic radiations (light). The goal of the paper on the other hand is to show that the wave-particle descriptions of reality can be applied to any physical situation simultaneously. The objective of the paper is to show that the Photoelectric Effect and the Compton Effect can both be explained by the wave model and the particle model at the same time.

Consider a long wire connected to an ammeter and strong electric and magnetic fields produced in a vacuum. Let us assume that whenever a wire is brought in vicinity of a changing electric field, electrons of mass (m) are set into motion in the wire and then an ammeter deflects, recording a current (i_E). The current in the wire due to a changing electric field should be given by

$$i_E = \frac{j\varepsilon_o}{2\pi m}$$

E (1)

Where (ε_o) is the permittivity of free space and (j) is the constant of action in SI units Js. therefore the current is quantized and depends on both the electric field and the mass of an electron.

When the wire is brought into the magnetic field, vibrating electrons at a frequency of oscillation (f) are set in motion at a speed (v) through the wire generating a current given by

$$i_B = \frac{v}{2\pi\mu_0 f} B$$

(2)

Where (μ_0) is the permeability of free space.

Assuming that the ammeter records different values of (i_E) and (i_B), what will be the change in the current values recorded at the ammeter? Subtracting equation (1) from equation (2) we have

$$\Delta I = (i_E - i_B) = \left(\frac{j\varepsilon_0}{2\pi m} E - \frac{v}{2\pi\mu_0 f} B\right)$$

(3)

This is the change in the currents due to changing magnetic and electric fields. Assuming that there is no change in the current, meaning that the current values for i_E are equal to those of i_B (i.e $\Delta I = 0$). This will imply that the magnetic field strength was equal to the electric field strength at one point in both experiments. In terms of electromagnetic radiations in the vacuum, assuming that a wire carrying current is replaced by a wave and electrons are replaced by photons. The wire replaced by a wave is made up of vibrating electric and magnetic fields at a given frequency making an electromagnetic wave. The electrons replaced with photons will represent the particle properties of the electromagnetic wave (light) with associated mass and speed (v).

The symmetry here is between the long wire and the wave, the electrons and the Photons. The electric and magnetic fields brought in vicinity of the wire and the number of oscillations per second of the electron in the wire is what leads to an electromagnetic wave. The electrons with a given mass and moving at a given speed is what constitute a photon. Then at $\Delta I = 0$, we have on arranging,

$$\frac{jf}{mv} = \frac{1}{2\pi\mu_0\varepsilon_0} \frac{B}{E}$$

(4)

This means that at $\Delta I = 0$, either a changing magnetic field or a changing electric field produces a current. Then it should be true that a changing magnetic field produces an electric field just as a changing electric field produces a magnetic field. This process in the electromagnetic wave continues indefinitely. The electromagnetic wave will move at a constant speed (c), since for electromagnetic waves, $\frac{E}{B} = c$, and for a photon $\frac{jf}{mv} = c$ where j=6.63× 10^{-34} Js (also called the Planck constant after Max Planck) and mv is the photon momentum. Implying that the photon energy is related to the frequency of the electromagnetic wave by (jf). Then the electromagnetic wave will move at a constant speed given as, since by symmetry $\frac{E}{B} = \frac{jf}{mv} = c$

$$c = \frac{1}{\sqrt{\varepsilon_o \mu_o}} = 2.99792458 \times 10^8 \frac{m}{s}$$

Where $\varepsilon_o = 8.85418782 \times 10^{-12} \frac{c^2}{Nm^2}$ and $\mu_o = 1.26 \times 10^{-6} \frac{Ns^2}{c^2}$

We have therefore deduced based on the symmetry between a current (electron) carrying wire in the electromagnetic field and the photons in electromagnetic waves that an electromagnetic wave moves at a constant speed of light. It is also true from the deductions that light is indeed made up of particles of light called photons and vibrating electric and magnetic fields. The deduction would not be possible if the wave and particle descriptions of the situations had not been applied simultaneously (into what is called "the wave-particle duality).

Unexpectedly enough the **photoelectric effect** can also be explained by Equation (3), on arranging

$$\frac{2\pi m f}{s_o E} \Delta I = jf - \frac{mv}{2\pi \mu_o s_o} \frac{B}{E}$$

Then the total energy of the particle of light (Photon) is then given by

$$jf = \frac{2\pi m f}{\varepsilon_0 E}\Delta I + \frac{mv}{2\pi \mu_0 \varepsilon_0}\frac{B}{E}$$

(5)

It is therefore true that the photoelectric effect can be explained when both the particle and wave models of reality are applied in the experiment at the same time (simultaneously). The work function from Einstein's photoelectric equation (A. Einstein, 1905) will here be replaced by $\frac{2\pi m f}{\varepsilon_0 E}\Delta I$ while the kinetic energy of the electrons at the surface of the metal will be given by $\frac{mv}{2\pi \mu_0 \varepsilon_0}\frac{B}{E}$. Equation (5) reduces to Einstein's Photoelectric effect when, the speed of the electron is $v = \frac{1}{\pi \mu_0 \varepsilon_0}\frac{B}{E}$ and the change in current for a complete circuit is $\Delta I = \frac{j\varepsilon_0 E}{2\pi m}$.

The validity of the **Compton Effect** can also be deduced from Equation (3). The current can be taken as the product of the frequency (f) of radiations and the charge (q) on the particle.

Then the current due to the electric field is $i_E = q f_1$ and that due to the magnetic field is $i_B = q f_2$. In the case of the Compton Effect, q is the charge on the free electron while f_1 and f_2 are the frequencies of the incoming photon and outgoing photon after collision with the free electron respectively. Then equation (3) can be written as

$$f_1 - f_2 = \frac{1}{q}\left(\frac{j\varepsilon_0}{2\pi m}E - \frac{v}{2\pi \mu_0 f}B\right)$$

(6)

Since photons move with the speed of light(c) then their frequencies is related to their speed and wavelength by $f = \frac{c}{\lambda}$, then we have

$$\frac{1}{\lambda_1} - \frac{1}{\lambda_2} = \frac{1}{qc}\left(\frac{j\varepsilon_0}{2\pi m}E - \frac{v}{2\pi \mu_0 f}B\right)$$

On arranging to include the charge density of the free electron for electric field lines in an area of $\frac{\lambda_1 \lambda_2}{2\pi}$, we obtain

$$\frac{2\pi q}{\lambda_1 \lambda_2}(\lambda_2 - \lambda_1) = \frac{j}{mc}\left(\varepsilon_o E - \frac{mv}{\mu_o jf}B\right)$$

Where (mc) is the momentum of an electron treated relativistic ally, on letting the charge density $\frac{2\pi q}{\lambda_1 \lambda_2} = \varepsilon_o E$, we deduce the change in the wave length of the incoming photon and outgoing photon after collision with the free electron as

$$\Delta\lambda = (\lambda_2 - \lambda_1) = \frac{j}{mc}\left(1 - \frac{mv}{\rho \mu_o jf}B\right)$$

Since $\rho = \varepsilon_o E$, we then have

$$\Delta\lambda = (\lambda_2 - \lambda_1) = \frac{j}{mc}\left(1 - \frac{\frac{mvB}{\varepsilon_o \mu_o E}}{jf}\right)$$

Since jf is the energy carried by the photon, and then also $\frac{mvB}{\varepsilon_o \mu_o E}$ is the energy carried by the free electron. Treating the electron relativistically such that for electromagnetic waves moving at a speed (v) relative to the electron moving at a speed of light $c = \frac{1}{\sqrt{\varepsilon_o \mu_o}}$, the electric field in the wave will be related to the magnetic field by $Bv = E$. then the energy carried by an electron can be given by mc^2. Then the angle at which the photon is scattered after collision with the free electron will be given by

$$\theta = \cos^{-1}\left(\frac{mvB}{\varepsilon_0\mu_0 E}\right)\Big/ jf$$

(7)

Where mv is the momentum of the photon in the electromagnetic wave consisting of a changing electric field E and magnetic field B both moving at a constant speed of light $c = \frac{1}{\sqrt{\varepsilon_0\mu_0}}$. Treating the electron relativistically we have

$$\theta = \cos^{-1}\frac{mc^2}{jf}$$

When the energy carried by the photon is equal to the energy possessed by the electron then $\theta = 0$, meaning that there is or there is no scattering and whatsoever there is no increase in photon wavelength hence $\Delta\lambda = 0$.

A complete theory of light can't fail to explain the structure of an atom. I therefore take a complete discussion of what goes on inside an atom only with the help of Bohr's energy levels which he derived using classical mechanics and quantum theory. Let $\Delta f = f_1 - f_2$ be an increase in the frequency of the electromagnetic radiations emitted from an atom. Then squaring both sides of equation (6) and arranging will give

$$4\pi^2\Delta f^2 = \frac{1}{m^2 q^2}\left(j\varepsilon_0 E - \frac{mv}{\mu_0 f}B\right)^2$$

$$4\pi^2 m^2 q^2\Delta f^2 = j^2\varepsilon_0^2 E^2 - 2\frac{j\varepsilon_0 EBmv}{\mu_0 f} + \frac{B^2 m^2 v^2}{\mu_0^2 f^2}$$

Dividing through by $64\pi^4 j^2\varepsilon_0^2$ and multiplying through by q^2 gives the energy of the atom as on arranging

$$\frac{mq^4}{16\pi^2 \pi^4 j^2 \varepsilon_0^2} = \frac{1}{64\pi^4 m\Delta f^2}\left((Eq)^2 - 2\frac{m(Eq)(Bqv)}{\mu_0 \varepsilon_0 (jf)} + \frac{(Bqv)^2 m^2}{\mu_0^2 \varepsilon_0^2 (jf)^2}\right)$$

The energy of the n-th level is since the reduced Planck constant is

$$n\hbar = \frac{nj}{2\pi}$$

$$\frac{mq^4}{32\pi^2 \pi^4 n^2 \hbar^2 \varepsilon_0^2} = \frac{1}{32\pi^2 n^2 m\Delta f^2}\left((Eq)^2 - 2\frac{m(Eq)(Bqv)}{\mu_0 \varepsilon_0 (jf)} + \frac{(Bqv)^2 m^2}{\mu_0^2 \varepsilon_0^2 (jf)^2}\right)$$

The expression on the left hand side of the equation is the quantized energy of an atom (Niels Bohr, 1913) while the right hand side of the equation represents the energy of the atom in terms of the forces associated with it. In the equation we let $H_e = Eq$ be the electric force for a particle moving in the electric field and $H_b = Bqv$, the magnetic force on a particle with charge q moving in the magnetic field. Since the speed of light is $= \frac{1}{\sqrt{\varepsilon_0 \mu_0}}$, then the quantized energy can be given as

$$W_n = \frac{1}{32\pi^2 n^2 m\Delta f^2}\left(H_e^2 - 2\frac{H_e H_b mc^2}{jf} + \frac{H_b^2 (mc^2)^2}{(jf)^2}\right)$$

Then on arranging we obtain

$$W_n = \frac{1}{32\pi^2 n^2 m\Delta f^2}\left(H_e - \frac{mc^2}{jf}H_b\right)^2$$

$$(8)$$

When the energy of an electron moving at a speed of light in atom is equal to the energy of the emitted photon, then

$$W_n = \frac{1}{32\pi^2 n^2 m\Delta f^2}(H_e - H_b)^2 = \frac{1}{32\pi^2 mn^2}\left(\frac{\Delta H}{\Delta f}\right)^2$$

$$(9)$$

Where $\Delta H = H_e - H_b$ is the difference or change between the electric force and the magnetic force in an atom, when the two forces balance (i.e. $H_e = H_b$), then $W_n = 0$ meaning that the total energy of an atom will cease to exist.

Therefore the total energy of an atom increases with the square of the change in the electric and magnetic forces which govern an electron but falls off as the square of the change in the frequency of the radiation emitted by it.

From equation (8) the ratio of the energy of an electron to that of the photon $\frac{mc^2}{if}$, is the limit at which if the energies are not equal you will not get a change in the electric and magnetic forces. Treating the ratio as a number $\tau = \frac{mc^2}{if}$, we get from equation (8)

$$W_n = \frac{1}{32\pi^2 mn^2}\left(\frac{H_e - \tau H_b}{f_1 - f_2}\right)^2$$

$$(10)$$

When $\tau = 0$, it means that the relativistic energy (mc^2) of an electron in an atom is zero, and that the total energy of an atom only increases with the electric force on the electron. The relationship (equation 10) is a complete expression for the laws according to which, by the theory here advanced, the structure of an atom should be viewed.

In conclusion, a complete theory of light is only possible if both the wave and particle descriptions of reality are applied to the physical situation at the same time. In discussing Young's double slit experiment for example we should be able with the formulas given above to treat the electromagnetic radiations on both a wave and particle model.

21 LOOP QUANTUM GRAVITY AND THE STRUCTURE OF SPACE, TIME, AND THE UNIVERSE

Quantum gravity is the field of theoretical physics[1] that tries to unify quantum mechanics[2] with general relativity. Quantum mechanics describes the three fundamental forces of nature[3] while general relativity[4] is a theory of the fourth fundamental force: gravity[5]. The goal everyone is waiting for to emerge from this unification is a "theory of everything[6]", or "Grand Unified Theory[7]" (GUT). In 1986[8], Abhay Ashtekar[9] reformulated Einstein's field equations of general relativity using what have come to be known as Ashtekar variables[10], a particular flavor of Einstein-Cartan theory[11] with a complex connection. He was able to quantize gravity using gauge field theory[12]. In the Ashtekar formulation, the fundamental objects are a rule for parallel transport[13] and a coordinate frame known as a vierbein[14] at each point. Because the Ashtekar formulation was background-independent, it was possible to use Wilson loops[15] as the basis for a nonperturbative quantization of gravity. Explicit

---

1. http://en.wikipedia.org/wiki/Theoretical_physics

2. http://en.wikipedia.org/wiki/Quantum_mechanics

3. http://en.wikipedia.org/wiki/Fundamental_interaction

4. http://en.wikipedia.org/wiki/General_relativity

5. http://en.wikipedia.org/wiki/Gravitation

6. http://en.wikipedia.org/wiki/Theory_of_everything

7. http://en.wikipedia.org/wiki/Grand_Unified_Theory

8. http://en.wikipedia.org/wiki/1986

9. http://en.wikipedia.org/wiki/Abhay_Ashtekar

10. http://en.wikipedia.org/wiki/Ashtekar_variables

11. http://en.wikipedia.org/wiki/Einstein-Cartan_theory

12. http://en.wikipedia.org/wiki/Gauge_field_theory

13. http://en.wikipedia.org/wiki/Parallel_transport

14. http://en.wikipedia.org/wiki/Vierbein

15. http://en.wikipedia.org/wiki/Wilson_loop

(spatial) diffeomorphism invariance of the vacuum state[16] plays an essential role in the regularization of the Wilson loop states.

Around 1990[17], Carlo Rovelli[18] and Lee Smolin[19] obtained an explicit basis of states of quantum geometry, which turned out to be labelled by Penrose's spin networks[20]. In this context, spin networks arose as a generalization of Wilson loops necessary to deal with mutually intersecting loops. Mathematically, spin networks are related to group representation theory and can be used to construct knot invariants[21] such as the Jones Polynomial.

The need for this chapter is to understand those problems involving the combination of very large mass or energy and very small dimensions of space, such as the behavior of black holes[22], and the origin of the universe[23]

The formula for the quantization of quantum gravity

The model is based on separating the gravitational field into the sum of two components; that is the background and the quantum field. The background left is one for all our calculations. But because loop gravity ignores the back ground space as a lost entity that does not occur in space, there fore the need to reconstruct quantum field theory from scratch without a background space is taken into account. I therefore suggest that the calculation should be performed by summing all possible space-times.

Quantum field theory[24] depends on particle fields embedded in the flat space-time of special relativity[25]. General relativity[26] models gravity as a

16. http://en.wikipedia.org/wiki/Vacuum_state

17. http://en.wikipedia.org/wiki/1990

18. http://en.wikipedia.org/wiki/Carlo_Rovelli

19. http://en.wikipedia.org/wiki/Lee_Smolin

20. http://en.wikipedia.org/wiki/Spin_network

21. http://en.wikipedia.org/wiki/Knot_invariant

22. http://en.wikipedia.org/wiki/Black_hole

23. http://en.wikipedia.org/wiki/Big_Bang

24. http://en.wikipedia.org/wiki/Quantum_field_theory

25. http://en.wikipedia.org/wiki/Special_relativity

curvature within space-time[27] that changes as a gravitational mass (m) moves. Assuming a spherical symmetric object that space time is of dimensions increasing from 1, 2, 3, 4...N, where N is the nth term of the dimensions. To quantize space and time is to create a space in which all of physics is quantized. The nature of the curved space surface is described by increasing powers in the Schwarzschild radius $R_s = Gm/c^2$, Hence describing the dimensions of space. Quantum mechanics explains the existence of discrete energy states in an atom, in away that the angular momentum of the atom must be quantized, which is also the case for quantum gravity. The equation for the quantization of the loop quantum gravity can then be written as,

$$\eta R_s + \beta R_s^2 + \mu R_s^4 + \ldots\ldots\ldots\ldots + \delta R_s^N = n\hbar \ [1]$$

Where $\eta = \sqrt{Beh}$, is the momentum of a particle probing another form of quantum mechanics, $\hbar = h/2\pi$, where h is Planck constant, $\beta = 8\pi Be$, e is the elementary charge, B is the magnetic field and finally $\mu = 256\pi^3 P/c^2$, where P is the intensity and c is the constant speed of light.

The energy equation

What changes is the form of the equation given above the rest remains constant. The principle behind this is that eqn1 can be changed to any form simply for purposes of calculating complex phenomenon. The energy to which we are concerned here is expressed as a general expression describing the energy scales forming smaller and larger matter entities in the universe. The energy will thus be given by,

$$\eta c + \beta c R_s + \mu c R_s^3 + \ldots\ldots\ldots\ldots + \delta c R_s^{N-1} = n\hbar c/R_s \ [2]$$

Note: the background space described by the Schwarzschild radius has changed, thus the above equation in any case can be used to calculate the basic properties of Black holes. Remember the Schwarzschild radius is the radius for a given mass where, if that mass could be compressed to fit within that

---

26. http://en.wikipedia.org/wiki/General_relativity

27. http://en.wikipedia.org/wiki/Spacetime

radius, no known force or degeneracy pressure could stop it from continuing to collapse into a gravitational singularity[28].

The mass equation

Having explored the energy scale we now form general equation that describes well the mass scale. This is also done the same way as eqn2 and therefore generate,

$$\eta/c + \beta R_s/c + \mu R_s{}^3/c + \ldots\ldots + \delta R_s{}^{N-1}/c = n\hbar/cR_s \ [3]$$

The maximal magnetic field

Assuming that the energy $W = \beta cR_s$ from eqn2 is equal to the energy $W = mc^2$, we hence obtain the magnetic field as,

$$B = c^3/8\pi Ge = 1.0054 \times 10^{53} \ N/Am.$$

Using this magnetic field in the energy equation, $W = \eta c$ we get the energy in the form $W = (c^2/2) \sqrt{\hbar c/G}$ where the quantity $\sqrt{\hbar c/G}$ is the Planck mass M_P at an energy of 6.119×10^{18} GeV.

Time taken by a black hole to evaporate and its entropy

The energy required here is given in Eqn2, it is at this, that the intensity $P = W/A\Delta t$, (where A is the area and t is the time) is used. We take the energy $W = \mu cR_s{}^3$ (from Eq2) as our interest from which we obtain the time as $\Delta t = 256\pi^3 R_s{}^3/Ac$. But with black holes the area will become exactly equal to the square of the Planck length as $A \sim L^2{}_p = \hbar G/8\pi c^3$ hence the change in time is given by $\Delta t = 63500.86\pi \ G^3 m^3/\hbar c^4$.

For entropy we set the energy to kT, where k is Stefan's-Boltzmann's constant and T is the temperature of the body. Now for $kT = \mu cR_s{}^3$, since Δt is known the entropy is thus given by $S = W/T = 78.96Ak \ c^3/ \pi \hbar G \sim A/4$. In

---

28. http://en.wikipedia.org/wiki/Gravitational_singularity

conclusion we state that the entropy of a black hole is proportional to the area of the event horizon.

The quantum Hall Effect

For this effect the momentum η is used. From Eqn2 we set, $\eta c = n\hbar / R_s$ which gives the magnetic flux as $4\pi R_s^2 B = nh/e$, from which the resistance is given by $\zeta = 4\pi R_s^2 B /e = nh/e^2$. for n= 1,2,3,4 the resistance is of a value 25833.8Ω

Maximum Intensity

Using eqn3 in this case, since B is known and P got from $\mu R_s^4 = n\hbar$; as $P = \hbar c^2/256\pi^3 R_s^4$, we hence obtain, $M_p /2 + m + M_p /m = M_p /m$, which gives $M_p + 2m = 0$, and for identical mass M =0, which is true. The intensity at the planck length that is for $R_s = L_p$ is $P = c^8/\pi\hbar G^2$

22MY APPROACH TO SEMI-CLASSICAL GRAVITY

For the past thirteen years, I have been working on the most important theories of physics from scratch without employing the methods of general relativity and quantum field theories and I have come up with promising results. I have deduced the Black hole thermodynamics from first principles, I have deduced the Wiedmann Franz law from scratch, the Stefan Boltzmann power law, The result for the earliest period of time in the history of the Universe, I have related the Chandrasker theory of white dwarfs with the Bohr theory of the Hydrogen atom-the results are suprising, the rest is history. This book gives a clear account of these fields of physics.

The truth is, I hate Einstein and Hawking. I don't like them because I find it hard to use their mathematical ideas to deduce the theories I desire. It was that hard for me to classify where in the scientific community I fall, at first I thought that my ideas where into the quantum gravity field section but this was a lie. The quantum theory of gravity has not been fully settled. It was yesterday that I realized that my ideas fell into the Semi Classical physical regime when I browsed it online;

"Semi-classical physics refers to a theory in which one part of a system is described quantum-mechanically whereas the other is treated classically" In general, it incorporates a development in powers of Planck's constant, resulting in the classical physics of powers 0, and the first nontrivial approximation into the powers of -1. (Wikipedia)

I am sorry, Semi-classical physics hasn't gained much interest, there are too many criticism about its meaning, researches into the field have been discouraged, few physicists have written about it and it is that unimportant. But anyway I am an amateur to venture into a field that is irrelevant. I don't give a damn what you think.

My first insight into the field of Semi- classical physics is traced back in 2010 in my first paper I published on arXiv.org titled "A hypothetical investigation into the realm of the microscopic and macroscopic universes beyond the standard model" This paper clearly shows that I was into the field without knowing. For sure I thought I was dealing with the field of Quantum Gravity by then.

Well, if you don't understand Semi-classical physics, Amateurs do. Below I show you why I think I understand the field and you surely do. I provide many ideas which I think the entire scientific community must investigate.

(a) The meaning of semi-classical physics to an amateur

Assuming an experiment where the classical electric force f_e is balanced over the classical gravitational force f_g to determine their strength, the result will show that, the ratio of the two forces will follow a power law in powers of n of the gravitational coupling constant as,

$$\frac{f_e}{f_g} = \alpha_g{}^n$$

The left hand side of the equation represents the classical part of the system while the right hand side represents the quantum mechanical part of the system.

Let the classical part be described by two constants;

G-The Universal gravitational constant

c- The constant speed of light

Into (G, c)

Let the Quantum mechanical part be described by two constants,

$\hbar$- The reduced Planck constant

c-The constant speed of light

Into ($\hbar$, c)

Then from the above assumption Semi-classical physics will reduce results combining the constants above into $(G, c, \hbar)$

From the above formula we can deduce the time and length units of measure formulas to help us understand the field better,

$$t_n = \frac{Gm}{c^5} \alpha_g^{-n}$$

Time

$$l_n = \frac{Gm}{c^2} \alpha_g^{-n}$$

Length

Where m denotes the mass of a particle or body and $\alpha_g = \frac{Gm^2}{\hbar c}$ is the gravitational coupling constant. You can also assume interactions involving the electromagnetic coupling constant. The different fields of physics resulting from the above classification for different powers of n from 0, 1, 2 and -1/2 are given below,

For n=0

Classical General Relativity

$$t_0 = \frac{Gm}{c^3}$$

$$l_0 = \frac{Gm}{c^2}$$

For n=1

Quantum mechanics

$$t_1 = \frac{\hbar}{mc^2}$$

$$l_1 = \frac{\hbar}{mc}$$

For n=2

Semi-classical gravity

$$t_2 = \frac{\hbar^2}{Gcm^3}$$

$$l_2 = \frac{\hbar^2}{Gm^3}$$

For n= -1/2

Planck Units

$$t_{-1/2} = \left(\frac{G\hbar}{c^5}\right)^{1/2}$$

$$l_{-1/2} = \left(\frac{G\hbar}{c^3}\right)^{1/2}$$

The above derivation gives out a clear description of Semi-classical physics to a lay person.

One can decide to use n as a spatial dimension of space.

(b) Applications of semi-classical physics

(i)Radiation intensity of a black hole

The classical part of a system

Let the classical total force on an electron in orbit at a distance r from the nucleus of an atom be related to its electromagnetic and gravitational forces by,

$$f = \frac{F_G F_e}{F_B}$$

Where F_G is the gravitational force, F_e is the electric force and $F_B = Bev$ is the magnetic force

The angular momentum of an electron is given classically as,

$$L = \frac{Gm^2}{c} = mvr$$

The Quantum mechanical part of the system

The angular momentum is quantized as,

$$L = \frac{K_e e^2}{c} = \hbar$$

On eliminating the constant speed of light c from both the expression of the angular momentums we have

$$mvr = \frac{F_G}{F_e}\hbar$$

The ratio $\frac{F_G}{F_e}$ represents the classical part of the system while $\hbar$ represents the quantum part.

Eliminating F_G from the above expression we get the magnetic power as,

$$F_B c = \frac{2\pi r^2 \lambda m v F_e^2}{h^2}$$

But the de Brogile wave length of an electron is $\lambda = h/mv$ and the surface area of the sphere of orbit of an electron is $A = 4\pi r^2$. Then the electromagnetic Intensity is given as,

$$I = \frac{F_B c}{A} = \frac{F_e^2}{2h}$$

Thus the intensity of a wave is proportional to the square of the electric force If we let the power of the electromagnetic wave be P= F_Bc, and n be the fine structure constant $\alpha = ke^2/\hbar c$, then the equation for the intensity of the classical electromagnetic wave comes out clearly as,

$$P = EB/\mu o = \quad 2\varepsilon_o E^2 c \quad ,$$

Where μo is the permeability of free space

Assuming the coupling of the forces to be,

$$\frac{f_e}{f_g} = a_g{}^n$$

Then at n = -1, and $f_g = \frac{c^4}{8\pi G}$ we have the electric force as,

$$f_e = \frac{\hbar c^5}{8\pi G^2 m^2}$$

Then the intensity of the radiations will be given as

$$I = \frac{f_e^2}{2h} = \frac{\hbar c^{10}}{256\,\pi^2 G^4 m^4}$$

This expression comes from treating the particle classically in one part and then quantum mechanically in another part. It can be clearly seen above, that we haven't used the mathematics of general relativity or quantum field theory to reach at the result.

(ii) The earliest period of time in the history of the universe

Classical part of the system

Let the acceleration due to gravity of a particle (say an electron) in the gravitational field be given as

$$g = R/\Delta t^2$$

Where is Δt the time and R is the distance of the particle from the source. If the particle radiates energy then the energy per unit time is,

$$P = c^5/G$$

Quantum mechanical part of the system

The power and time must be quantized in units of $\hbar = h/2\pi$ where h is Planck constant, hence

$$P\Delta t^2 = n^2\hbar$$

Where n= 1,2,3........ is the principle quantum number.

But the potential energy of the electron in the various energy states is,

$$W = -ke^2/R$$

where k is the Coulomb constant and e is the elementary charge. Since Δt^2 is known from the expression for acceleration due to gravity. Then the distance R is,

$$R = n^2 Gg\,\hbar\,/c^5$$

From which the total energy is given by,

$$W = -\,ke^2c^5/n^2Gg\,\hbar$$

From the Bohr-Einstein frequency (f) condition, applied to a transition from a level with n = n_i to a level with n = n_f, The energy of a photon emitted by a hydrogen atom is given by the difference of two hydrogen energy levels

$$hf = E_i - E_f$$

Since frequency $f = c/\lambda$, where λ is the wavelength.Then we have,

$$1/\lambda = [ke^2c^4/2\pi G\hbar^2][1/g][1/n_f^2 - 1/n_i^2]$$

The equation obtained above shows some how a great significance of gravity in the quantum theory. So far it states that regardless of the levels in the transitions of an atom the acceleration due to gravity of the particles in the atom do greatly affect the nature of its spectrum.

The quantity $[ke^2c^4/2\pi G\hbar^2]$ in the formula above is the inverse of the square of time t and therefore ,

$$1/t^2 = [ke^2c^4 / 2\pi G\hbar^2]$$

From which the time is obtained as t = 1.58873 $\times 10^{-42}$s. This is the earliest period of time in the history of the universe.

(iii) The Weidmann Franz- Lorenz law

Treating one part of the system classically (macroscopic) and the other quantum mechanically (microscopic), we have the formula for the electric force acting on an electron in motion as

$$F = \frac{n^2}{a_g} f_g$$

Where n, is the principle quantum number.

The above formula differs from the one previously given. On squaring the above equation we obtain the square of the electric field as,

$$E^2 = \frac{n^4 c^4}{G^2 e^2 m^2} \left(\frac{c^3 \hbar}{8\pi G m} \right)^2$$

From the formula for the temperature of the black hole, the function $\frac{c^3 \hbar}{8\pi G m}$ is related to temperature as kT, and then the law for thermal conductivity will be reduced as,

$$\frac{\pi^2 E^2 G^2 m^2}{3 T c^4} = \left(\frac{n^4 \pi^2}{3} \right) \left(\frac{k}{e} \right)^2 T$$

The left hand side represents the ratio of the thermal conductivity K to the electric conductivity δ. The right hand side is the Weidman −Franz law. Therefore the left side of the equation represents the macroscopic part of the system while the right hand side represents the microscopic part of the system. Then the left-hand side will be given as,

$$\frac{K}{\delta} = \frac{1}{3}\left(\frac{\pi Gm}{c^2}\right)^2 \frac{E^2}{T} \quad \frac{\pi A}{3}\frac{E^2}{T}$$
$$=$$

Where A is the surface area of a body $A = \pi r_s^2$ with the schwarzichild's radius r_s. This is the conductivity ratio of a black hole.

PARTT V IN CLOSING

23 AN EXCEPTIONALLY SIMPLE QUANTUM THEORY OF GRAVITY

A precise and consistent quantum theory of gravity has not yet been proved, not even by the self proclaimed geniuses of this time. We are aware and satisfied that classical General Relativity is the most precise description of gravity due to its predictable nature. The left hand side of Einstein field equation represents the metric of space time curvature while the right hand side represents the matter - energy content of the classical matter fields of pressure and energy density. It is known that quantum mechanics plays an important role in the behaviour of the matter fields but has no place in the Einsteins field equations. According to S.W.Hawking (1975), one therefore has a problem of defining a consistent scheme in which the space time metric is treated classically but is coupled to the matter fields which are treated quantum mechanically.

In this book we propose that, in order to estimate stellar parameters to a high degree of accuracy for both microscopic and macroscopic descriptions of white dwarfs and black holes one has to treat the right hand side of Einstein field equation quantum mechanically as,

$$\left(\frac{8\pi G}{c^3}\right)^{3/2} \frac{m_H}{\hbar^{1/2}} P_{eg}$$

$P_{eg} = \frac{f_e f_g}{\hbar c}$, where P_{eg} is the total pressure, f_g is the gravitational force, f_e is the electric force , G is the gravitational constant, c is the constant speed of light, $\hbar$ is the reduced planck constant and m_H is the mass of an Hydrogen atom.

Proof of the Chandrasker Mass Limit and the Lowest Principal Quantum Number from a New Approach to Quantum Gravity

Although in the Bohr theory of an hydogen atom orbit quantization doesnot permit a lower orbit than the bohr radius of $a_0 = 0.53\text{Å}$, this section sets out to show that this is not the case with white dwarfs due to the state of a hydrogen atom under high pressure.

We know from the Chandrasker derivations that, the equation governing the hydrostatic equilbrium of a star is given by

$$- r^2 P(r) = GM(r)\rho(r) \qquad \frac{dP}{dr} = -\frac{GM(r)}{r^2}\rho$$

o r

Where P denotes the total pressure, ρ is density, and M(r) is the mass interior to a sphere of radius r.

We could however write the same equation in a different form given by

$$P_{eg}r^2 = \frac{\hbar^{1/2}c^{9/2}}{(8\pi G)^{3/2}m_H}, \text{ where } P_{eg} = \frac{f_e f_g}{\hbar c} \ (1)$$

1.1 The total Gravitational Binding Energy of a Star

The electric potential energy $E_e = f_e r$ as we know it can be can be deduced from (1) and is given by,

$$E_e = \left(\frac{\hbar}{8\pi G}\right)^{3/2} \frac{c^{11/2}}{E_g m_H}$$

where E_g is the gravitational potential binding energy given by $f_g r$

Using the principle of energy equipartition, we assume that the electric binding energy is of order the discrete energy of an hydrogen atom from Bohrs theory as,

$$E_n = E_e \Rightarrow \left(\frac{\hbar}{8\pi G}\right)^{3/2} \frac{c^{11/2}}{E_g m_H} = \frac{m_H k_e{}^2 e^4}{2n^2 \hbar^2}$$

where, k_e is the Coulomb constant, e is the charge on an electron and n is the principal quantum number.

From the above assumptions the gravitational binding energy is given as,

$$E_g = 2\left(\frac{\hbar c}{8\pi G}\right)^{3/2} \left(\frac{nc}{\alpha_e m_H}\right)^2$$

(2)

where α_e is the fine structure constant $\frac{ke^2}{\hbar c} = 1/137$

In Table 1 we list the values of E_g for several values of n-the principal quantum number. From this table it follows in particular, that the higher the principal quantum number, the higher the gravitational binding energy of a star.

Table 1.

The total gravitational binding energy of a star

n(Principal quantum number)	E_g(Joules)	Remarks
0.003212	3.229×10^{47}	
0.0345	3.468×10^{48}	White Dwarf
1	1.005×10^{50}	

What do we conclude from the foregoing calculation? We conclude that equation (2) is at the base of the equilbrium of actual stars in relation to the energy state and binding energy of the Hydrogen atom. It differs from the Chandrasker calculation by the introduction of a natural fine structure constant, providing the energy of proper magnitude for the measurement of stellar energies and therefore proving to be a better theory for stellar structure.

This could be elaborated in detail by flowers original words,

"The Black-dwarf material is best likened to a single gigantic molecule in its lowest quantum state. On the Fermi-Dirac statistics, its high density can be achieved in one and only one way, in virtue of a correspondingly great energy content. But this energy can no more be expended in radiation than the energy of a normal atom or molecule. The only difference between Black-dwarf matter and a normal molecule is that the molecule can exist in afree state while the black dwarf matter can only so exist under high external pressure.

1.2 The Theory of White -Dwarf Stars and Black Holes; The Limiting Mass at the Lowest Principal Quantum Number

The gravitational energy is known to be of order $E_g = \dfrac{GM^2}{r}$, M being the mass of a star. Then equating this to equation (2) we obtain the radius of a star as,

$$r = \frac{1}{2}\left(\frac{8\pi G}{\hbar c}\right)^{3/2}\left(\frac{\alpha_e m_H}{nc}\right)^2 GM^2$$

(3)

while the above equation states that the radius is proportional to the square of it's mass, the Chandrasker analysis is in disagreement, stating that r is inversely proportional to the cube root of the mass.

But at a point where r equation (3) approaches the schwarzichilds radius r_s

$$r \Rightarrow r_s \ \frac{1}{2}\left(\frac{8\pi G}{\hbar c}\right)^{3/2}\left(\frac{\alpha_e m_H}{nc}\right)^2 GM^2 = \frac{2GM}{c^2}$$

,

We obtain an upper limit to the mass of,

$$M = 4\left(\frac{n}{\alpha_e}\right)^2\left(\frac{\hbar c}{8\pi G}\right)^{3/2}\frac{1}{M_H{}^2}$$

(4)

Now consider equating the original solution of Chandrasker mass limit to our newly developed formula (4) , we have

$$M_C = M(4), \ \frac{\omega^0{}_3\sqrt{3\pi}}{2}\left(\frac{\hbar c}{8\pi G}\right)^{3/2}\frac{1}{\mu_e{}^2 M_H{}^2} = 4\left(\frac{n}{\alpha_e}\right)^2\left(\frac{\hbar c}{8\pi G}\right)^{3/2}\frac{1}{M_H{}^2}$$

$$n = \frac{\alpha_e}{\mu_e}\left(\frac{\omega^0{}_3}{8}\sqrt{3\pi}\right)^{1/2}$$

$\omega^0{}_3 = 2.018236$, is a constant connected with the solution to the lane-Emden equation, and $\mu_e = 2$, average molecular weight per electron,then

$$n = 3.212 \times 10^{-3}$$

In Table 2 we list the values of M and r for several values of n-the principal quantum number, including the one calculated above.

Table 2.

The Mass limit and radius limit of a star

n(Principal quantum number)	M (Kilograms)	r (meters)	Remarks
3.212×10^{-3}	2.304×10^{28}	34.153	$(0.012 M_{sun}$
0.0345	2.66×10^{30}	3944.601	Chandrasekar mass limit $(1.4 M_{sun}$
1	2.234×10^{33}	3.31×10^6	Maximum mass of a white dwarf

What do we conclude from the foregoing calculation? We conclude that the formation of a white dwarf star or any other stellar structure will never exceed the Schwarzichild's radius of 34.153m, this will only happen at the most lowest quantum principal number of 3.212×10^{-3}. For example, at the principal quantum number the size of the fine structure constant $\frac{1}{137}$, the mass obtained will be of $0.063 M_{sun}$ and r=176.443m. Therefore under high external pressure the minimum mass of a last star that is formed is of order $2.304 \times 10^{28} kg$ and this only occurs at r=34.153m under the lowest energy state below the known Bohrs radius of $a_0 = 0.53\text{Å}$.

What is Wrong With Hawking Temperature

In his paper "Particle creation by Black holes" Hawking pointed out that "In the classical theory black holes can only absorb and not emit particles. However it is shown that quantum mechanical effects cause black holes to create and emit particles as if they were hot bodies with temperature $\frac{\hbar c^3}{8\pi GMk} \approx 10^{-6}\left(\frac{M_{sun}}{M}\right)^0 K$,". However this is not the case when the assumptions given in the first sections of this book are taken into account. For example, we know that, the electric potential energy is given by,

$$E_e = \left(\frac{\hbar}{8\pi G}\right)^{3/2} \frac{c^{11/2}}{E_g m_H}$$

But treating the particles in the process General relativisticaly (at the Schwarzichild radius), the gravitational potential energy will be of order $E_g = \frac{mc^2}{2}$, giving the electric energy as,

$$E_e = 2\left(\frac{\hbar}{8\pi G}\right)^{3/2} \frac{c^{7/2}}{mm_H}$$

Now the thermal energy is given by $E_{thermal} = kT$, where k is the Boltzmann constant.

By the principal of Equipartition

$$E_{thermal} \sim E_e \Rightarrow T = 2\left(\frac{\hbar}{8\pi G}\right)^{\frac{3}{2}} \frac{c^{\frac{7}{2}}}{kmm_H}, \qquad (5)$$

$$T = 3.3891 \times 10^{11} \frac{M_{sun}}{M}$$

NOTE: IN A LIMIT WHERE m_H IS THE PLANCK MASS $m_H = \sqrt{\frac{\hbar c}{8\pi G}}$, EQUATION 5 ABOVE FOR THE TEMPERATURE OF A BLACK

HOLE REDUCES TO THE HAWKING TEMPERATURE FORMULA

$$T = \frac{\hbar c^3}{8\pi GMk}$$

For conditions at the centre of the Sun, $T = 3.3891 \times 10^{11}K$ which is in disagreement with the Hawking temperature of $T_H = 6.476 \times 10^{-8}K$. This is left for the reader to analyse.

2.1 Entropy of a Black Hole

For derivations which i will not show here, I am led to the total energy of a Black Hole given by,

$$E_B = \frac{2A(\hbar c^{13})^{1/2}}{(8\pi G)^{5/2}m_H m}$$

where, A is the surface area of the event horizon

But since the entropy is energy per unit temperature,

$$S = \frac{E_B}{T}$$

Remember that temperature is given by equation(5),

Then the entropy will be given by,

$$S = \frac{Ac^3 k}{4\pi G\hbar}$$

This is in agreement with the Bekenstein-Hawking area entopy law

First Hand Accounts/ Insights into the Theory Given Above

3.1 On the Development of a Quantum Gravity-Hydrostatic Equation and its Implication to Physics

It is known that the equation governing the hydrostatic equilibrium of a star is given by,

$$\frac{dP}{dr} = -\frac{GM(r)}{r^2}\rho$$

$$(1)$$

Where P denotes the total pressure, ρ is density, and M(r) is the mass interior to a sphere of radius r.

what if we rewrite the above formula in a form given by,

$$\frac{F_g F_e}{\hbar c}r^2 = \frac{c^4}{8\pi G} = constant$$

$$(2)$$

where F_g is the gravitational force, F_e is the electric force, c is the speed of light, $\hbar$ is the reduced Planck constant and G is the gravitational constant.

let the pressure be, $P = \frac{F_g F_e}{\hbar c}$, this means that pressure is dependent on the product of the gravitational and electric forces in a quantum-relativistic realm.

therefore in simple terms, we can write (2) in its simplest form as, $\frac{dP}{dr} = -\frac{c^4}{8\pi G r^3}(r)$

to include the density, we have

$$\frac{dP}{dr} = -\frac{rc^2}{8\pi Gr_s}(r)\rho$$

where r_s is schwarzichild's radius. thus at $r_s = r$, the star will form a black hole.

To differ from (1) we have formulated one of the first quantum gravity -hydrostatic equation.

from (2) we can write the electric potential energy as,

$$F_e r = \frac{\hbar c^5}{8\pi G E_g}$$

where, $E_g = F_g r$ is the gravitational potential energy

at a point where the potential gravitational energy is in equilibrium with the potential electric energy the total energy is that of the Planck energy by,

$$E = F_e r = E_g = \sqrt{\frac{\hbar c^5}{8\pi G}} = 3.91 \times 10^8 J$$

since the Bohr energy of an hydrogen atom is given by,

$$E_n = \frac{mk_e^2 e^4}{2n^2 \hbar^2}$$

then, using the principle of equipartition of energy

$$F_e r = E_n$$

we deduce ,the gravitational potential energy as

$$E_g = \frac{2n^2 \hbar^3 c^5}{8\pi G m_e k_e^2 e^4}$$

this can be written in a simplest form as,

$$E_g = \frac{2n^2}{m_e}\left(\frac{J_p}{\alpha_e}\right)^2$$

(3)

where, $J_p = \sqrt{\frac{\hbar c^3}{8\pi G}}$ is the planck momentum 1.3035N.s

$\alpha_e = \frac{ke^2}{\hbar c}$ is the fine structure constant $\frac{1}{137}$

m_e is the mass of an electron $9.11 \times 10^{-31} kg$

then the total gravitational energy is calculated to be,

$E_g = 7.00 \times 10^{34} J(n^2)$

for a thermal energy kT, we estimate a temperature of

$$T = \frac{7.00 \times 10^{34}}{k} = 5.07 \times 10^{57} K$$

we know that, the gravitational potential energy is given by, $\frac{GM^2}{r}$,

for,

$$E_g = \frac{GM^2}{r} = \frac{2n^2}{m_e}\left(\frac{J_p}{\alpha_e}\right)^2$$

then the radius mass relation can be written as,

$$r = \frac{GM^2 m_e}{2n^2}\left(\frac{\alpha_e}{J_p}\right)^2$$

$$r = 9.527 \times 10^{-46}\frac{M^2}{n^2}$$

for the solar mass $M = 1.9 \times 10^{30} kg$, $r \sim 3.44 \times 10^{15} m$

if r is equal to the schwarzichilds radius $\frac{GM}{c^2}$,

then,

$$\frac{GM}{c^2} = \frac{GM^2 m_e}{2n^2} \left(\frac{\alpha_e}{J_p}\right)^2$$

the solar mass limit is given by

$$M = \frac{2n^2}{m_e c^2} \left(\frac{J_p}{\alpha_e}\right)^2 \sim 7.78 \times 10^{17} kg$$

but for $r = \frac{2\pi\hbar}{m_e c}$ compton wavelength

then

$$\frac{2\pi\hbar}{m_e c} = \frac{GM^2 m_e}{2n^2} \left(\frac{\alpha_e}{J_p}\right)^2$$

from which mass reduces to,

$$M = \frac{2n\hbar^{1/2}\pi^{1/2}}{G^{1/2} m_e c} \frac{J_p}{\alpha_e} \sim 2.913 \times 10^{12} kg$$

This is the minimum mass of a Black hole

Derivation of the Temperature and Entropy for Hawking Radiations of a Black Hole Using Simple Semi- Classical Quantum Gravity Approaches

The mathematical ideas behind the theory of general relativity and quantum mechanics are complicated and thus difficult to grasp in all perspectives and this could be one of the reasons why it has been and is still difficult to prove a complete quantum theory of gravity for most physicist. It is in the same domain that I also find the derivations behind the Hawking radiations in literature tiresome and at the same time incomplete. This is why I find it relevant to derive the temperature and entropy and other relations of Black hole physics from first principles, which I think will shed light on the future of quantum gravity and a requirement for the re-development of quantum mechanics and general relativity approaches in a much simpler way as I have labored to deduce it here under three approaches.

4.1 The First Approach

Analogous to Einstein's field equation in the theory of General Relativity we create a formula where the scalar curvature is related to the classical Newtonian gravitational and electric forces by,

$$\frac{1}{R^2} = S\frac{F_g F_e}{\hbar c}$$

1)

Where R is the radius of curvature

F_g is the gravitational force

F_g is the electric force

$\hbar$ is the reduced planck constant

c is the constant speed of light

$S = \dfrac{8\pi G}{c^4}$, G is the Newton's gravitational constant

From the above expression it is seen that, the expression on the right is the pressure and energy density of matter treated quantum relativisticaly in units of $\hbar c$, as, Pressure = energy density= $\dfrac{F_g F_e}{\hbar c}$.

Temperature

Let the work done by the electric force to move a particle through a distance R be given by,

$$W = F_e R$$

But the electric force can be got from equation 1, hence giving the work done as,

$$W = \dfrac{\hbar c}{s F_g R}$$

and the Newtonian force can be given by, $F_g = \dfrac{Gm^2}{R^2}$ hence deducing the work done as,

$$W = \dfrac{\hbar c R}{s G m^2}$$

For a black hole the size of the schwarzichild's radius $R = \dfrac{Gm}{c^2}$, we obtain

$$W = \frac{\hbar}{smc}$$

Assuming that the work done is analogous to the thermal kinetic energy of an ideal gas, W=kT, where, k is the Boltzmann's constant and T is the temperature of the particle, then the temperature of the radiated energy or radiation from the black hole will be given as,

$T = \frac{\hbar}{smck}$ or, $T = \frac{\lambda}{sk}$, where $\lambda = \frac{\hbar}{mc}$ is the deBrogile-wavelength of a particle moving at the spaeed of light. Thus the temperature of the radiated energy from the black holes increases with the wavelength of the emitted particle.

Entropy

Let the entropy be, $S = \frac{E}{T}$, and let the total energy radiated E be given by,

$$E = \frac{F_e^2}{4\pi\hbar} tA$$

Where, t is the time taken by a black hole to dissipate and A is the surface area of the black hole event horizon.

But since the electric force is known from equation 1, then

$$E = \frac{\hbar c^2}{4\pi S^2 R^4 F_g^2} tA$$

Let time be, $t = \frac{mc}{F_e} = \frac{msR^2F_g}{\hbar}$

Then the total energy will be given by

$$E = \frac{mc^2}{4\pi sR^2 F_g} A$$

Since $F_g = \frac{Gm^2}{R^2}$

Then, $E = \frac{1}{4\pi sR_s} A$, where R_s is the schwarzichild's radius $R_s = \frac{Gm}{c^2}$

Then entropy can be deduced since T is known as, $T = \frac{\lambda}{sk}$, then dividing E by T we obtain,

$$S = \frac{Ak}{4\pi R_s \lambda}$$

If a black hole was a particle with mass m, we can easily compute it's radius and it's wavelength, combining the two, one can compute a new surface area as $A_s = 4\pi R_s \lambda$, then the entropy of a black hole will be given by,

$$S = \frac{A}{A_s} k ,$$

If we let the thermodynamic probability be W, then the probability for work done in the expansion from A_s to A is, $W = e^{\frac{A}{A_s}}$.

4.2 Second Approach

We write a set of formulas from which our derivations will proceed

1) It is well known that the electric field is force per unit charge but here a generalized equation for an electric field created by an electron exhibiting wave properties in the nucleus of an atom in the gravitational field on a quantum scale is given by

$$E = \frac{1}{r}\sqrt{\frac{Gm^3f}{2\hbar\varepsilon_o}} \quad 2$$

Then the electric force in this case will be formulated as

$$F_1 = \frac{e}{r}\sqrt{\frac{Gm^3f}{2\hbar\varepsilon_o}} \quad 3$$

2) The surface area at a radius r of orbit of an electron of mass m around the nucleus of an atom in a wave like manner is given by

$$\text{surface area}(A) = \frac{4\mu_o e^2}{m} \quad 4$$

3) The time taken by the magnetic field B of an electron to pass through a given surface is

$$\text{time}(t) = \frac{\lambda\varepsilon_o AB}{e} \quad 5$$

Note: the above expression is the same as Faraday's induction law.

4) The gravitational force acting on all matter in the universe or the modified gravitational force is given as

$$F_2 = \left(\frac{Gm^3}{r^2}\right)\left(\frac{e}{2B\lambda\hbar\varepsilon_o}\right) \quad 6$$

The above formulas are important in deriving the formula for the temperature, entropy and the time taken by a black hole to evaporate as shown below;

Temperature of a Black Hole

It is known that the kinetic energy KE of molecules in the Boltzmann hypothesis is related to the temperature of the body in question in this case a black hole (in relation to the black body) by $KE = \varphi T$ where φ is Boltzmann's constant. The formula for the kinetic energy can be derived by using a hypothesis that the electromagnetic force – coulombs force is equal to eqn3 as

$$\frac{ke^2}{r^2} = \frac{e}{r}\sqrt{\frac{Gm^3 f}{2\hbar\varepsilon_0}}$$

On squaring both sides of the equation, cancelling like terms and taking into account that the frequency of an electron is $f = \frac{v}{\lambda}$, then the kinetic energy of an electron inside the black hole is given by

$$KE = \frac{\lambda\mu_0 e^2}{A}\frac{c^3\hbar}{8\pi Gm^2}$$

Since the surface area is given as from eqaution4 then the kinetic energy of molecules or particles (for an ideal gas) within the black hole will be given by

$$KE = \frac{c^3\hbar}{8\pi Gm} = T\varphi_7$$

Then from Boltzmann's relationship the temperature of the black hole is formulated as

$$T = \frac{c^3\hbar}{8\pi Gm\varphi}8$$

The Entropy of the Black Hole

By definition entropy is a measure of disorder. To solve the entropy of black holes we shall consider a very complex argument about the entropy in question. We assume that the modified gravitational force given by equation6 is identical to the modified electric field given by equation3 as, $\left(\frac{Gm^3}{r^2}\right)\left(\frac{e}{2B\lambda\hbar\varepsilon_0}\right) \equiv \frac{e}{r}\sqrt{\frac{Gm^3 f}{2\hbar\varepsilon_0}}$ in otherwise the two forces are equal but opposite. Then squaring both sides of the equation and multiplying through by Gc^5 one obtains a new relation of forces on both sides given as

$$\frac{tc^7}{16\pi G^2 m} = \frac{Ac^6}{32\pi rmG^2}$$

Both the left and right hand side represent a force. From the left hand side t is the expression of time given by $t = \frac{\hbar e^2}{2m^3 c^2 G\varepsilon_0}$. Note: the left hand side force is the pull of matter inside the black hole while the right hand side force is the force acting on particles or matter at the surface of the black hole.

Since the heat is the product of the force on a particle and the distance r from the centre of the black hole, then using the force on the right hand side of the above equation the heat will be given by

$$Q = \frac{Ac^6}{32\pi mG^2}$$

Remember the temperature of the black hole is also known from equation6 and by definition the entropy of the system is the change in heat per unit temperature $\frac{Q}{T}$, then the entropy of the black hole will be given by

$$S = \frac{A\varphi c^3}{4G\hbar} \quad 9$$

This implies that the entropy of a black hole is proportional to its surface area.

The Time Taken by a Black Hole to Evaporate

Assuming that particles that formed a black hole are moving away or are separating from it after a given time of its existence, if we measure the relative speed of these particles in relation to the energy they carry we obtain a relationship given by

$$\frac{v^2}{c^2} = \frac{8\pi G}{c^2}\left(\frac{W}{8\pi r}\right)_{10}$$

Where v is the velocity of these particles as measured relative to the speed of light c and W is the energy carried by the particles as they move away from the centre of the black hole at a distance r.

If we let the force causing the particles to separate from the black hole be given as $\frac{Gm^3 e\ v}{2r\lambda B\hbar\varepsilon_o c}$, then the energy of these particles will be given by

$$W = \frac{Gm^3 e\ v}{2r\lambda B\hbar\varepsilon_o c}$$

Substituting this in equation8, we obtain a relationship of time as given by the law 3 of equation 5 as

$$t = \frac{v^2}{c^2}\left(\frac{\pi G^2 m^3}{\hbar c^4}\right)$$

The velocity of the particles in the astronomical lab will be measured as v= 4.193E6 m/s and since the speed of light is a constant then the time taken by a black hole to evaporate is given by

$$t = \frac{5120\pi G^2 m^3}{\hbar c^4}$$

4.3 Third Approach

Temperature of a black hole

It is here by hypothesized that, the gravitational field will create particles and emit them only if the electromagnetic force of such particles were equal to the force (unknown in literature) $F = \frac{Me}{r}\sqrt{\frac{Gp}{2\hbar\varepsilon_o\lambda}}$ .Where p, is the momentum of a particle. under general conditions, the force given will reduce to the Reissner-Nordstrom metric as given here, if the momentum of an electron at a distance r from the singularity point to the event horizon is related to the de Brogile wavelength as $p = \frac{2\pi\hbar}{\lambda}$, and both the distance r and wavelength λ was the product of the speed of light c and the period T as r=cT and $\lambda = cT$, then the force will be given by $F = \frac{Mp}{rh}\sqrt{\frac{Ge^2}{4\pi\varepsilon_0}}$, but since $\frac{p}{2\pi\hbar} = \frac{1}{\lambda}$, then we have, $F = \frac{2\pi M}{T^2}\sqrt{\frac{Ge^2}{4\pi\varepsilon_0 c^4}}$, this reduces to $F = \frac{2\pi M}{T^2}r_q$, where $r_q = \sqrt{\frac{Ge^2}{4\pi\varepsilon_0 c^4}}$ is the Reissner-Nordstrom radius of a charged black hole.

Having derived the Reissner-Nordstrom metric from our force formula, we now return to our exercise of deriving the temperature of a black hole. We consider a particle with charge e, exhibiting deBrogile wave properties of momentum and wavelength from the centre of mass M of a black hole. We then assume that this particle experiences an electromagnetic force due to the magnetic and electric field created by other particles in its surrounding area. The same particle also experiences a force due to the strong gravitational field emanating from the black hole. Equating the two forces as $\frac{Me}{r}\sqrt{\frac{Gp}{2\hbar\varepsilon_0\lambda}} = \frac{e^2}{4\pi\varepsilon_0 r^2}$, from this expression we obtain the momentum of a particle as $p = \frac{\hbar e^2 \lambda}{2\pi A\varepsilon_0 GM^2}$. This is the momentum possessed by a particle (emitted by the gravitational field of a black hole) at the surface of the event horizon, where $A = 4\pi r^2$ is the spherical surface area of the horizon.

For relativistic effects, the kinetic energy of a particle will be related to its momentum by K.E=pc and to the Boltzmann's law by K.E=kT, where k is the Boltzmann's constant and T is the absolute temperature. By similarity we can equate the two energies as pc=kT, then from the equation of momentum we can obtain the temperature as,

$$T = \frac{\hbar e^2 \lambda c}{2\pi A\varepsilon_0 GM^2 k}.$$

Expressing the permittivity of free space in terms of the permeability of free space $\varepsilon_0 = \frac{1}{\mu_0 c^2}$, we obtain the Hawking temperature of a black hole as,

$$T = \left(\frac{4e^2\mu_0\lambda}{AM}\right)\frac{\hbar c^3}{8\pi GMk}$$

In a more general form, in terms of energies it can be expressed as,

$$T = \left(\frac{4e^2\lambda}{A\varepsilon_0 Mc^2}\right)\frac{\hbar c^3}{8\pi GMk}\,_{11}$$

We propose that, $mc^2 \geq \frac{4e^2\lambda}{A\varepsilon_0}$ and if $A = 4\pi R^2$ and $\lambda = \frac{R}{4}$ then, $mc^2 \geq \frac{e^2}{4\pi\varepsilon_0 R}$ the electric potential energy.

Entropy of a black hole

In an attempt to prevent the violation of the generalized second law of thermodynamics, Bekenstein proposed a universal upper bound on the ratio entropy to energy for bounded systems (Phys RevD23, 287-1981), which was later rejected by Unruh and Wald in 1982. They proposed a thought experiment in which a box lowered down into a black hole felt an effective buoyancy force which was caused by the acceleration radiation felt by the box near the black hole. They argued further that, this buoyancy force would guarantee a lower bound on the energy gain of the black hole, hence saving the generalized second law without a need for entropy bound.

In this section we give a formula for the buoyancy force which is different from the Unruh and Wald formula which appeared in their 1982 paper.

At a distance r from the center of mass m of a black hole, the buoyancy force is given by,

$$F_B = \frac{rc^6}{8G^3m}\,_{12}$$

From the above force formula the energy gain by the black hole will be given by,

$$W_B = \frac{Ac^6}{32\pi G^3 m}$$

Where, A is the area of the event horizon. Since entropy is the ratio of energy to temperature, $S_B = W_B/T_B$ and temperature of a black hole is known from equation 11, then the entropy of a black hole is given by,

$$S_B = \frac{Akc^3}{4G\hbar}\left(\frac{A\varepsilon_0 Mc^2}{4e^2\lambda}\right) \quad 13$$

Development Of a Simple Quantum Theory Of Gravity

5.1 An exceptionally simple quantum theory of gravity

The Einstein field equation is written in the form, $G_{\mu\nu} + \Lambda g_{\mu\nu} = \frac{8\pi G}{c^4} T_{\mu\nu}$ where, the expression on the left represents the curvature of space time while the expression on the right represents the matter-energy content of the universe. Then assuming a quantum state in which a gravitating particle of radius R is acted upon by all classical forces, the expression on the left, the metric of space time curvature can be written in a special form as,

$$G_{\mu\nu} + \Lambda g_{\mu\nu} = \frac{\hbar c}{F_u R^4} \quad \text{.................1}$$

Where, F_u is the force arising from a quantized field in zero point vacuum energy, $\hbar$ is the reduced planck constant and c is the constant speed of light.

Whereas the expression on the right (the stress-energy tensor) will be written in a form,

$$T_{\mu\nu} = \frac{F_G F_E}{F_u R^6} \quad \text{.........................2}$$

Where, F_G is the Newtonian classical gravitational force between two particles of mass m, F_E is the electrostatic force between two particles of charge q and G is the gravitational constant.

Then in terms of the pressure and energy density the stress-energy tensor is,

$$T_{\mu\nu} = P_g \varphi = \rho_E \varphi$$

Where $\varphi = \frac{F_E}{F_u}$ is the coupling of forces, P_g is the pressure ($P_g = \frac{GM^2}{R^4}$) and ρ_E is the energy density or the potential gravitational energy per unit volume R^3.

Then in its simple form, the Einstein Field equation may be expressed as,

$$\frac{hc}{F_E R^4} = \frac{8\pi G}{c^4} P_g = \frac{8\pi G}{c^4} \rho_E \quad \text{................3}$$

From the above equation, the gravitational potential field is analogous to the quantum gravitational potential by

$$\nabla^2 \phi = 4\pi G \rho_E (\text{Classical})$$

And, $\quad \nabla^2 \phi = \frac{hc^5}{2F_E R^4}$ (quantum)

We have coupled a quantum system to a classical one by simply denoting the metric of space time in Einstein's field equation as $\frac{hc}{F_u R^4}$, this out come gives us a unique technique through which we can express the gravitational effects in terms of quantum mechanics.

5.2 Einstein Field Equation for a Relationship between the DeBrogile Wavelength and the Energy Density of an Electromagnetic Wave

In a cyclotron the acceleration of a particle describing circular motion at a distance R in a magnetic field B will be given as

$$a_x = \frac{2\pi B v R}{q\mu_0} ,$$

Where μ_0 is the permeability of free space

q is the charge on a particle and

v is a velocity at right angles to the direction of the field B

But when quantum and gravitational effects are taken into account, we are led to a different formula for the acceleration given by

$$a_y = \frac{hc^5}{8\pi G m R^2 E_\zeta} ,$$

This is deduced from equation3 where, the electrostatic force on a charge q in vicinity of the electric field E is $F_E = Eq$, E=Bc and the inertial force is $F_G = ma_y$.

Then at a point where the two accelerations are equal that is, $a_x = a_y$, we are led to,

$$\frac{\lambda}{2\pi R^3} = \left(\frac{8\pi G}{c^4}\right)\frac{EB}{\mu_0 C} = \left(\frac{8\pi G}{c^4}\right)\rho$$

Where $\rho = \frac{EB}{\mu_0 C}$ is the energy density of an electromagnetic wave in vacuum and $\lambda = \frac{\hbar}{mv}$ is the deBrogile wave length

The formula obtained above is the solution to the Einstein field equation in which the wave properties of matter in terms of the DE Brogile wavelength are related to the wave properties of an electromagnetic wave in terms of the energy density of an electromagnetic wave. The expression on the left represents the quantum nature of wave mechanics while that on the right represents the classical nature of electromagnetic waves interrelated by the gravitational constant.

It is therefore true from our derivations that, when the classical acceleration of particles in the cyclotron is equal in magnitude to the modern acceleration (not yet observed), we deduce properties of a wave on both a quantum and classical realm simultaneously. This means that both the wave and particle properties of matter cannot be separated in any experiment and or observation, hence a wave –particle duality of matter

.

5.3 Derivation of the Schwarzschild-Hawking Power Law

Suppose a force F_E does work on a Black hole of mass M to move it through a small displacement Δd in time Δt, where $\Delta d / \Delta t$ is the average speed v, then the power is,

$$P = F_E v$$

But from equation 3, $F_E = \dfrac{\hbar c^5}{8\pi G^2 M^2}$, is the force required to displace or accelerate a black hole, the force increases as the mass of the black hole decreases.

If we let $v = c$ then the power of a Black hole will be given by,

$$P = \frac{\hbar c^6}{8\pi G^2 M^2}$$

This sets a limit to which velocity a black hole can be accelerated. Note: the above formula has not yet been derived in the frame work of semi classical gravity. If this is semi classical gravity, then we are towards achieving a quantum theory of gravity. I therefore leave the derivation above to the entire scientific community to investigate.

However the expression above differs from that deduced from the Stefan-Boltzmann radiation power law of

$$P = \frac{\hbar c^6}{15360\pi G^2 M^2}$$

.

Meaning that, it requires a velocity of $1.5625 \times 10^5 m/s$ to obtain this power from our derivations.

5.4 Derivation of the Bekenstein –Hawking Area Entropy Law

The energy or work done by a black hole will to a great degree depend on the surface area of the event horizon A and on the Compton wavelength λ of a black hole provided the force exerted on this black hole remains a constant as,

$$W = F_E \frac{A}{\lambda}$$

The Compton wavelength is $= \frac{2\pi\hbar}{mc}$, and F_E is known, thus the energy is,

$$W = \frac{Ac^6}{16\pi^2 G^3 m}$$

This implies that, the energy of a black hole is proportional to the surface area of the event horizon but inversely proportional to its mass.

If we apply the above statement to entropy which is energy per unit temperature, $S=W/T$ we can deduce the entropy of a black hole. Let us deduce the expression for temperature: when the electric force applied on a body of mass m through a schwarzichild's radius $R_s = \frac{2Gm}{c^2}$, results into an energy equal to the translational kinetic energy as, $F_E R_s = kT$, where k is the Boltzmann's constant. Then the expression for temperature will be given as, $T = \frac{\hbar c^3}{4\pi Gmk}$, this is the temperature of a black hole. Then substituting for the energy and temperature in the entropy formula we obtain,

$$S = \frac{Ac^3 k}{4\pi Gh}$$ 8

This is the entropy of a black hole in its simplest form.

In conclusion, the Book has presented a new approach to Quantum Gravity that is different from string theory and loop quantum gravity by Carlo Rovelli and Edward Witten. The major result of the research is the derivation of the Bekenstein-Hawking area entropy law from first principles using new methods with a well defined calculation where no infinities appear. As far as this book is concerned there is no other theory from which such a calculation can proceed. Hence the methods used in here are the only one from which a detailed quantum theory of gravity "Holy Grail of modern physics" precedes and where the result of the Bekenstein-Hawking area entropy law can be achieved.

ADDITIONAL READINGS

Balungi Francis, (2010) "A hypothetical investigation into the realm of the microscopic and macroscopic universes beyond the standard model" general physics arXiv:1002.2287v1[1] [physics.gen-ph]

Hawking, Stephen[2] (1975). "Particle Creation by Black Holes"[3]. Commun. Math. Phys.[4] 43 (3): 199–220. Bibcode[5]:1975CMaPh..43..199H[6].

Hawking, S. W.[7] (1974). "Black hole explosions?". Nature.248(5443):30–31.

Bibcode[8]:1974Natur.248...30H[9].doi[10]:10.1038/248030a0[11].

Carlo Rovelli (2003) "Quantum Gravity" Draft of the Book Pdf

Some few texts used are from Wikipedia https://creativecommons.org/licenses/by-sa/3.0/

D. N. Page, Phys. Rev. D 13, 198 (1976).

C. Gao and Y.Lu, Pulsations of a black hole in LQG (2012) arXiv:1706.08009v3

A.H. Chamseddine and V.Mukhanov, Non singular black hole (2016) arXiv 1612.05861v1

---

1. https://arxiv.org/abs/1002.2287v1

2. https://en.wikipedia.org/wiki/Stephen_Hawking

3. http://www.springerlink.com/content/c4553033029k5wk6/

4. https://en.wikipedia.org/wiki/Commun._Math._Phys.

5. https://en.wikipedia.org/wiki/Bibcode

6. http://adsabs.harvard.edu/abs/1975CMaPh..43..199H

7. https://en.wikipedia.org/wiki/Stephen_Hawking

8. https://en.wikipedia.org/wiki/Bibcode

9. http://adsabs.harvard.edu/abs/1974Natur.248...30H

10. https://en.wikipedia.org/wiki/Digital_object_identifier

11. https://doi.org/10.1038%2F248030a0

M.Bojowald and G.M.Paily, A no-singularity scenario in LQG (2012) arXiv: 1206.5765v1

P.Singh, class.Quant.Grav,26,125005(2009), arXiv:0901.2750

P.Singh and F.Vidotto, Phys.Rev, D83,064027(2011) arXiv:1012.1307

C.Rovelli and F.Vidotto, Phy. Rev,111(9) 091303(2013) arXiv:1307.3228v2

M.Bojowald, Initial conditions for a universe, Gravity Research Foundation (2003)

A.Ashtekar, Singularity Resolution in Loop Quantum Cosmology (2008) arXiv:0812.4703v1

J.Brunneumann and T.Thiemann, On singularity avoidance in Loop Quantum Gravity (2005) arXiv:0505032v1

L.Modesto, Disappearence of the Black hole singularity in Quantum gravity (2004) arXiv:0407097v2

Mikhailov, A.A. (1959).Mon. Not. Roy. Astron. Soc.,119, 593.

P. Merat etal.(1974). Astron & Astrophys 32, 471-475

Trempler, R.J. (1956).Helv. Phys. Acta, Suppl.,IV, 106.

Trempler, R.J. (1932). " The deflection of light in the sun's gravitational field "Astronomical society of the pacific 167

Einstein, A. (1916).Ann. d. Phys.,49, 769; (1923).The Principle of Relativity, (translators Perret, W. and Jeffery, G.B.), (Dover Publications, Inc., New York), pp. 109–164.

Von Klüber, H. (1960). InVistas in Astronomy, Vol. 3, pp. 47–77.

K. Hentschel (1992). Erwin Finlay Freundlich and testing Einstein theory of relativity, Communicated by J.D. North

Muhleman, D.O., Ekers, R.D. and Fomalont, E.B. (1970).Phys. Rev. Lett.,24, 1377

Mikhailov, A.A. (1956).Astron. Zh.,33, 912.

Dyson, F.W., Eddington, A.S. and Davidson, C. (1920).Phil. Trans. Roy. sog., A220, 291

Chant, C.A. and Young, R.K. (1924).Publ. Dom. Astron. Obs.,2, 275.

Campbell, W.W. and Trumbler, R.J. (1923).Lick Obs. Bull.,11, 41.

Freundlich, E.F., von Klüber, H. and von Brunn, A. (1931).Abhandl. Preuss. Akad. Wiss. Berlin, Phys. Math. Kl., No.l;Z. Astrophys.,3, 171

Mikhailov, A.A. (1949).Expeditions to Observe the Total Solar Eclipse of 21 September, 1941, (report), (ed. Fesenkov, V.G.), (Publications of the Academy of Sciences, U.S.S.R.), pp. 337–351.

S.P. Martin, in Perspectives on Supersymmetry , edited by G.L. Kane (World Scientific, Singapore, 1998) pp. 1–98; and a longer archive version in hep-ph/ 9709356; I.J.R. Aitchison, hep-ph/0505105.

Mara Beller, Quantum Dialogue: The Making of a Revolution. University of Chicago Press, Chicago, 2001.

Morrison, Philp: "The Neutrino, scientific American, Vol 194,no.1 (1956),pp.58-68.

R. Haag, J. T. Lopuszanski and M. Sohnius, Nucl. Phys. B88, 257 (1975) S.R. Coleman and J. Mandula, Phys.Rev. 159 (1967) 1251.

H.P. Nilles, Phys. Reports 110, 1 (1984).

P. Nath, R. Arnowitt, and A.H. Chamseddine, Applied N = 1 Supergravity (World Scientific, Singapore, 1984).

S.P. Martin, in Perspectives on Supersymmetry , edited by G.L. Kane (World Scientific, Singapore, 1998) pp. 1–98; and a longer archive version in hep-ph/ 9709356; I.J.R. Aitchison, hep-ph/0505105.

S. Weinberg, The Quantum Theory of Fields, VolumeIII: Supersymmetry (Cambridge University Press, Cambridge,UK, 2000).

E. Witten, Nucl. Phys. B188, 513 (1981).

S. Dimopoulos and H. Georgi, Nucl. Phys. B193, 150(1981).

N. Sakai, Z. Phys. C11, 153 (1981);R.K. Kaul, Phys. Lett. 109B, 19 (1982).

L. Susskind, Phys. Reports 104, 181 (1984).

L. Girardello and M. Grisaru, Nucl. Phys. B194, 65(1982); L.J. Hall and L. Randall,

Phys. Rev. Lett. 65, 2939(1990);I. Jack and D.R.T. Jones, Phys. Lett. B457, 101 (1999).

For a review, see N. Polonsky, Supersymmetry: Structureand phenomena. Extensions of the standard model, Lect.Notes Phys. M68, 1 (2001).

G. Bertone, D. Hooper and J. Silk, Phys. Reports 405, 279 (2005).

G. Jungman, M. Kamionkowski, and K. Griest, Phys. Reports 267, 195 (1996).

V. Agrawal, S.M. Barr, J.F. Donoghue and D. Seckel,Phys. Rev. D57, 5480 (1998).

N. Arkani-Hamed and S. Dimopoulos, JHEP 0506, 073(2005); G.F. Giudice and A. Romanino, Nucl. Phys. B699, 65(2004) [erratum: B706, 65 (2005)]. July 27, 2006 11:28

en.wikipedia.org/wiki/Supersymmetry - 52k - Cached[12] - Similar pages[13]

en.wikipedia.org/wiki/Grand_unification_theory - 39k - Cached[14] - Similar pages[15]

12. http://64.233.169.104/search?q=cache:ZBSZWNLrdxEJ:en.wikipedia.org/wiki/
    Supersymmetry+supersymmetry&hl=en&ct=clnk&cd=1&gl=ug

13. http://www.google.co.ug/search?hl=en&q=related:en.wikipedia.org/wiki/Supersymmetry

14. http://64.233.169.104/search?q=cache:Le3KZW7QnUYJ:en.wikipedia.org/wiki/
    Grand_unification_theory+grand+unification&hl=en&ct=clnk&cd=1&gl=ug

In cosmology, the Planck epoch (or Planck era), named after Max Planck, is the earliest period of time in the history of the universe, en.wikipedia.org/wiki/**Planck_epoch** - 23k - Cached[16] - Similar pages

L. Shapiro and J. Sol`a, Phys. Lett. B 530, 10 (2002);

E. V.Gorbar and I. L. Shapiro, JHEP 02, 021 (2003); A. M. Pelinson,

L. Shapiro, and F. I. Takakura, Nucl. Phys. B 648, 417 (2003).

A. Starobinsky, Phys. Lett. B 91, 99 (1980).

G. F. R. Ellis, J. Murugan, and C. G. Tsagas, Class. Quant. Grav.21, 233 (2004).

H. V. Peiris et al., Astrophys. J. Suppl. 148, 213 (2003).

D. N. Spergel et al., astro-ph/0603449.

Vilenkin, Phys. Rev. D 32, 2511 (1985).

A. Starobinsky, Pis'ma Astron. Zh 9, 579 (1983).

A.H. Guth, Phys. Rev. D23, 347 (1981).

A.D. Linde, Phys. Lett. B108, 389 (1982); A. Albrecht, P.J.

Steinhardt, Phys.Rev. Lett. 48, 1220 (1982).

A.D. Linde, Phys Lett. B129, 177 (1983).

N. Makino, M. Sasaki, Prog. Theor. Phys. 86, 103 (1991);

D. Kaiser, Phys. Rev.D52, 4295 (1995).

H. Goldberg, Phys. Rev. Lett. 50, 1419 (1983).

E. Kolb and M. Turner, The Early Universe (Addison-Wesley, Reading, MA,1990).

---

15. http://www.google.co.ug/search?hl=en&q=related:en.wikipedia.org/wiki/Grand_unification_theory

16. http://64.233.169.104/search?q=cache:d5zWrem4T08J:en.wikipedia.org/wiki/

   Planck_epoch+planck+epoch&hl=en&ct=clnk&cd=1

W. Garretson and E. Carlson, Phys. Lett. B 315, 232(1993); H. Goldberg, hep-ph/0003197.

Eddington, A. S., The Internal Constitution of the Stars (Cambridge University Press, England,1926), p. 16

Duncan R .C. & Thompson C., Ap.J.392, L 9 (1992).

Thompson , C, Duncan , R .C ., Woods , P., Kouveliotou , C ., Finger , M.H. & van Parad ij s , J .,ApJ, submitted , astro-ph /9908086, (2000).

Schwinger , J .,Phys. Rev.73, 416L (1948)

Carlip, S.: Quantum gravity: a progress report. Rept. Prog. Phys. 64, 885 (2001).arXiv:gr-qc/0108040

Kerr,R.P.: Gravitational field of a spinning mass as an example of algebraically special metrics.

Phys. Rev. Lett. 11, 237–238 (1963)

Bekenstein, J.: Black holes and the second law. Lett. Nuovo Cim. 4, 737–740 (1972)

Bardeen, J.M., Carter, B., Hawking, S.: The four laws of black hole mechanics. Commun.

Math. Phys. 31, 161–170 (1973)

Tolman, R.: Relativity, Thermodynamics, and Cosmology. Dover Books on Physics Series.

Dover Publications, New York (1987)

Oppenheimer, J., Volkoff, G.: On massive neutron cores. Phys. Rev. 55, 374–381 (1939)

Tolman, R.C.: Static solutions of einstein's field equations for spheres of fluid, Phys. Rev. 55,364–373 (1939)

Zel'dovich Y.B.: Zh. Eksp. Teoret. Fiz.41, 1609 (1961)

Bondi, H.: Massive spheres in general relativity. Proc. Roy. Soc. Lond. A281, 303–317 (1964)

Sorkin, R.D., Wald, R.M., Zhang, Z.J.: Entropy of selfgravitating radiation. Gen. Rel. Grav. 1127–1146 (1981)

Newman, E.T., Couch, R., Chinnapared, K., Exton, A., Prakash, A., et al.: Metric of a rotating,charged mass. J. Math. Phys. 6, 918–919 (1965)

Ginzburg, V., Ozernoi, L.: Sov. Phys. JETP 20, 689 (1965)

Doroshkevich, A., Zel'dovich, Y., Novikov I.: Gravitational collapse of nonsymmetric and rotating masses, JETP 49 (1965)

Israel, W.: Event horizons in static vacuum space-times. Phys. Rev. 164, 1776–1779 (1967)

Israel,W.: Event horizons in static electrovac space-times. Commun. Math. Phys. 8, 245–260 (1968)

Loop quantum gravity does provide such a prediction [363, 364], and it disagrees with the semiclassical

Carter, B.: Axisymmetric black hole has only two degrees of freedom. Phys. Rev. Lett. 26, 331–333(1971)

Penrose, R.: Gravitational collapse: the role of general relativity. Riv. Nuovo Cim. 1, 252–276 (1969)

Christodoulou, D.: Reversible and irreversible transformations in black hole physics. Phys. Rev. Lett. 25, 1596–1597 (1970)

Christodoulou, D., Ruffini, R.: Reversible transformations of a charged black hole. Phys. Rev. D4, 3552–3555 (1971)

Hawking, S.: Particle creation by black holes. Commun. Math. Phys. 43, 199–220 (1975)

Klein, O.: Die reflexion von elektronen an einem potential sprung nach der relativistischen dynamik von dirac. Z. Phys. 53, 157 (1929)

Wald, R.M.: General Relativity. The University of Chicago Press, Chicago (1984)

Hawking, S.W.: Black hole explosions. Nature 248, 30–31 (1974)

Hawking, S., Ellis, G.: The large scale structure of space-time. Cambridge University Press, Cambridge (1973)

Carter, B.: Black hole equilibrium states, In Black Holes—Les astres occlus. Gordon and Breach Science Publishers, (1973)

Hawking, S.W.: Gravitational radiation from colliding black holes. Phys. Rev. Lett. 26, 1344– 1346 (1971)

Hawking, S.: Black holes in general relativity. Commun. Math. Phys. 25, 152–166 (1972)

Bekenstein, J.: Extraction of energy and charge from a black hole. Phys. Rev. D7, 949–953 (1973)

Bekenstein, J.D.: Black holes and entropy. Phys. Rev. D7, 2333–2346 (1973)

Hawking, S.: Quantum gravity and path integrals. Phys. Rev. D18, 1747–1753 (1978)

Gross, D.J., Perry, M.J., Yaffe, L.G.: Instability of flat space at finite temperature. Phys. Rev. D25, 330–355 (1982)

Unruh, W.G., Wald, R.M.: What happens when an accelerating observer detects a rindler particle. Phys. Rev. D29, 1047–1056 (1984)

Bekenstein, J.D.: Auniversal upper bound on the entropy to energy ratio for bounded systems. Phys. Rev. D23, 287 (1981)

Unruh,W.,Wald, R.M.: Acceleration radiation and generalized second law of thermodynamics. Phys. Rev. D25, 942–958 (1982)

Unruh, W., Wald, R.M.: Entropy bounds, acceleration radiation, and the generalized second law. Phys. Rev. D27, 2271–2276 (1983)

Image : MPI for gravitational physics / W.Benger-zib

Tomilin,K.A., (1999). "Natural Systems Of Units: To The Centenary Aniniversary Of The Planck Systems", 287-296

Sivaram, C. (2007). "What Is Special About the Planck Mass"? arXiv:0707.0058v1

H. Georgi and S.L. Glahow. (1974) "Unity Of All Elementary-Particle Forces". Phys. Rev. Letters 32, 438

INDEX

"The important thing in science is not so much to obtain new facts as to discover new ways of thinking about them"

-William Bragg

SUSP Science Foundation- presents answers to the major unsolved problems in physics from a fresh mind.

Questions you will encounter .....and perhaps even solve;

Problem of time

Cosmic inflation

Horizon problem

Origin of the universe

Size of the universe

Baryon asymmetry

Cosmological constant problem

Dark matter

Dark energy

Dark flow

Axis of evil

Shape of the universe

Hierarchy problem

Planck particle

Magnetic monopoles

Proton decay

Supersymetry

Generation of matter

Neutrino mass

Colour confinement

Diffuse interstellar bands

Yang-mills theory

Physical information

Supernovae

p-nuclei

ultra-high energy cosmic ray

color confinement

vacuum catastrophe

quantum gravity

black holes

information paradox

extra dimensions

cosmic censorship hypothesis

locality

strong cp problem

axions

anomalous magnetic moment

proton radius puzzel

penta quarks

exotic hadrons

mu problem

koide formula

astrophysical jet

supermassive black holes

galaxy rotation problem

large scale anisotropy

space roar

ultra luminous pulsar

origin of mass

fast radio bursts

quantum chromodynamics

bose einstein condensation

theory of everything

quantum field theory

Don't miss out!

Visit the website below and you can sign up to receive emails whenever Balungi Francis publishes a new book. There's no charge and no obligation.

https://books2read.com/r/B-A-LFLG-NJOZ

BOOKS 2 READ

Connecting independent readers to independent writers.

Also by Balungi Francis

Beyond Einstein
Quantum Gravity in a Nutshell1
Solutions to the Unsolved Physics Problems
Mathematical Foundation of the Quantum Theory of Gravity
A New Approach to Quantum Gravity
Balungi's Approach to Quantum Gravity
QG: The strange theory of Space,Time and Matter
The Holy Grail of Modern Physics
Fifty Formulas that Changed the World
Quantum Gravity in a Nutshell1 Second Edition
What is Real?:Space Time Singularities or Quantum Black Holes?Dark
Matter or Planck Mass Particles? General Relativity or Quantum Gravity?
Volume or Area Entropy Law?
The Holy Grail of Modern Physics
Brief Solutions to the Big Problems in Physics, Astrophysics and Cosmology

Brief Solutions to the Big Problems
Brief Solutions to the Big Problems

Pursuing a Nobel Prize
Serious Scientific Answers to Millennium Physics Questions

Using Geographical Information Systems to Create an Agroclimatic Zone map for Soroti District

Think Physics
Proof of the Proton Radius
Emergence of Gravity
On the Deflection of Light in the Sun's Gravitational Field
Reinventing Gravity

Standalone
Using Gis to Create an Agroclimatic Zone map for Soroti Distric
Expecting
Quantum Gravity in a Nutshell 2
Balungi's Guide to a Healthy Pregnancy
Prove Physics
The Origin of Gravity and the Laws of Physics
Derivation of Newton's Law of Gravitation
When Gravity Breaks Down